# SILFS

## Volume 4

# Current Topics in Logic and the Philosophy of Science

Papers from SILFS 2022 postgraduate conference

Volume 1
New Essays in Logic and Philosophy of Science
Marcello D'Agostino, Guilio Giorello, Federico Laudisa, Telmo Pievani and Corrado Sinigaglia, eds.

Volume 2
Open Problems in Philosophy of Sciences
Pierluigi Graziani, Luca Guzzardi and Massimo Sangoi, eds.

Volume 3
New Directions in Logic and the Philosophy of Science
Laura Felline, Antonio Ledda, Francesco Paoli and Emanuele Rossanese, eds.

Volume 4
Current Topics in Logic and the Philosophy of Science. Papers from SILFS 2022 postgraduate conference
Francesco Bianchini, Vincenzo Fano and Pierluigi Graziani, eds.

**SILFS Series Editor**
Marcello D'Agostino                    marcello.dagostino@unimi.it

# Current Topics in Logic and the Philosophy of Science

Papers from SILFS 2022 postgraduate conference

Edited by

Francesco Bianchini

Vincenzo Fano

Pierluigi Graziani

© Individual author and College Publications 2024. All rights reserved.

ISBN 978-1-84890-455-2

College Publications
Scientific Director: Dov Gabbay
Managing Director: Jane Spurr
Department of Computer Science

http://www.collegepublications.co.uk

Original cover design by Laraine Welch

---

All rights reserved. No part of this publication may be reproduced, stored in a retrieval system or transmitted in any form, or by any means, electronic, mechanical, photocopying, recording or otherwise without prior permission, in writing, from the publisher.

# CONTENTS

## ARTICLES

Introduction .................................................. 1
    *Francesco Bianchini, Vincenzo Fano, Pierluigi Graziani*

Polysemy in Entropic Model Selection for Deterministic Finite
    Automata Learning ......................................... 5
    *John Fergus William Smiles*

Feynman's Theorizing and Visualization ............... 21
    *Marco Forgione*

Realism, Underdetermination, and Inference
    in Cognitive Neuroscience ................................. 37
    *Davide Coraci, Gustavo Cevolani*

Can People Unlearn? A Reflection on the Conceptual and
    Cognitive Foundations of Organizations Systems Theory . 55
    *Samuele Maccioni, Cristiano Ghiringhelli, Edoardo Datteri*

The Problem of Time for Non-Deparametrizable Models and
    Quantum Gravity .......................................... 75
    *Álvaro Mozota Frauca*

Artificial Intelligence as Expected Intelligence . . . . . . . . . . .  89
  *Martina Bacaro, Francesco Bianchini*

Framing Beliefs into Fractional Semantics for Classical Logic  117
  *Matteo Bizzarri*

Ignorance and Its Formal Limits  . . . . . . . . . . . . . . . . . . 131
  *Ekaterina Kubyshkina, Mattia Petrolo*

A Note on Schematicity and Completeness in Prawitz . . . . . 143
  *Antonio Piccolomini d'Aragona*

The Use of Experts in Probabilistic Seismic Hazard Analysis:
  Towards a Confidence Approach . . . . . . . . . . . . . . . . . 159
  *Luca Zanetti, Daniele Chiffi, Lorenza Petrini*

# INTRODUCTION

FRANCESCO BIANCHINI*, VINCENZO FANO**
PIERLUIGI GRAZIANI**
*University of Bologna; ** University of Urbino
francesco.bianchini@unibo.it; vincenzo.fano@uniurb.it;
pierluigi.graziani@uniurb.it

It may seem that logic and the philosophy of sciences are well-formed disciplinary fields and not subject to change but only to enrichment. They are not scientific disciplines in a narrow sense, they are not "sciences" in the proper, empirical sense of the term, at least in their philosophical interpretation, which is largely theoretical. Therefore, unlike the empirical sciences, one can think that the accumulation of knowledge, which once characterized the empirical disciplines as the main concept and was then abandoned for epistemological reasons, is valid. Logic and the philosophy of sciences would instead be only accumulative. However, this is not the case. As this volume demonstrates, logic and the philosophy of sciences are disciplinary fields that change their underlying themes, broaden the investigated aspects and principles, and address interdisciplinary and transdisciplinary areas that are constantly born and reborn. The selection of papers stemming from the 2022 SILFS Post-Graduate Conference and which constitute the contributions of this volume appears as a trace of this expansion in new directions, all the more significant because it derives from "new," in the sense of young, scholars who grasp the fundamental aspects of these fields of study and push them in contemporary directions, broadening their boundaries. This comprehensive volume combines, therefore, diverse cutting-edge research across multiple scientific domains. The selected articles offer in-depth explorations of various themes, providing readers with a rich tapestry of

---

We thank the *Italian Society for Logic and the Philosophy of Science* for supporting the conference from which these selected papers originate.

insights into contemporary scientific endeavors. Let's delve into each article, analyzing their unique contributions to the volume.

**John Fergus William Smiles**, with *Polysemy in Entropic Model Selection for Deterministic Finite Automata Learning*, discusses learning grammars from data, emphasizing complexity and data compression techniques. It introduces polysemy, highlighting its challenge in deterministic finite automata (DFA). The conclusion underscores the inherent complexity of inductive inference tasks.

**Marco Forgione**, in *Feynman's Theorizing and Visualization*, delves into the realm of quantum phenomena, exploring how Feynman's visualization skills shaped his understanding of quantum phenomena, emphasizing their role in narrative explanations and scientific method, with historical examples supporting this perspective.

**Davide Coraci** and **Gustavo Cevolani**, in *Realism, underdetermination, and inference in cognitive neuroscience*, discuss realism/antirealism in cognitive neuroscience, focusing on fMR"s reverse inference issue. It argues for a qualified realist stance, highlighting parallels between neuroscientific and philosophical discussions on empirical underdetermination.

**Samule Maccioni**, **Cristiano Ghiringhelli**, and **Edoardo Datteri**, in *Can People Unlearn? A Reflection on the Conceptual and Cognitive Foundations of Organizations Systems Theory* discuss organizations as organisms and emphasize continuous improvement and unlearning for survival. It explores the concept's evolution, addresses gaps, and aims to provide a clearer, empirically testable definition. The work contributes to understanding and informs future research on organizational unlearning.

**Alvaro Mozota Frauca**'s text *The problem of time for non deparametrizable models and quantum gravity* critiques canonical quantization in quantum gravity, focusing on the problem of time. General relativity's non-deparametrizability makes the problem and the issues at the time of solving it far more worrisome.

**Martina Bacaro** and **Francesco Bianchini**, in *Artificial Intelligence as Expected Intelligence* examine measuring intelligence in artificial intelligence (AI), emphasizing expected intelligence aligned with human interaction needs. It explores AI development, focusing on robotics and proposing parameterizable characteristics. Conclusions highlight future investigation prospects.

**Matteo Bizzarri**, in *Framing beliefs into fractional semantics for classical logic*, introduces fractional semantics, and presents fractional semantics for classical logic, extending $GS4$. It introduces $GS4_B$ for incorporating deductively closed beliefs, resulting in values exceeding traditional fractional semantics.

**Ekaterina Kubyshkina** and **Mattia Petrolo**, in *Ignorance and Its Formal Limits* highlight challenges in representing ignorance in epistemic logic, proposing diverse frameworks for nuanced distinctions beyond standard non-knowledge models.

**Antonio Piccolomini d'Aragona**, in *A note on schematicity and completeness in Prawitz* explores Prawitz's semantics, focusing on SVA and Base Semantics, highlighting intuitionistic incompleteness and the nuanced transition between them.

**Luca Zanetti**, **Daniele Chiffi**, and **Lorenza Petrini**, in *The Use of Experts in Probabilistic Seismic Hazard Analysis: Towards a Confidence Approach*, explore seismic hazard analysis, discussing aleatoric and epistemic uncertainties. It proposes a confidence-based approach, emphasizing scientists' input for better hazard estimates.

All these themes, which in many of the covered aspects are frontier, show not only the richness of the involved disciplinary fields, both in logic and in the philosophy of science, but also often bring with them a disciplinary interpenetration in which all this knowledge, and the theoretical skills connected to it, are intertwined in a concrete and abstract dimension at the same time. Therefore, as can be seen from the contributions, philosophy, and physics cannot do without logic and mathematics, epistemology is enriched in different ways with formal contents, technological and cognitive disciplines find the exact and extensive conceptual dimension in their philosophical framework. And the connections and permutations don't end there.

As we conclude this introductory journey, an invitation is extended to the readers – not merely to peruse these pages but to engage with the profound insights offered actively. Immerse yourselves in the intellectual richness of these articles, traverse the realms of scientific inquiry laid bare within these pages, and embark on a journey of discovery that transcends the confines of this volume. May the readers find inspiration in the intricate tapestry woven by these scholarly contributions, and may this exploration catalyze further intellectual pursuits. Happy reading, and may pursuing knowledge be as rewarding as the discoveries that await these pages!

# Polysemy in Entropic Model Selection for Deterministic Finite Automata Learning

### J.F.W. Smiles
*University of Bristol*
zg21169@bristol.ac.uk

**Abstract.** This chapter addresses the challenge of unsupervised grammar learning, focusing on the inductive inference problem of generalising from observed data to unseen instances. It explores the methodology of learning through data compression and the Minimum Description Length (MDL) principle, which optimises the data description to achieve a balance between model simplicity and accuracy. The chapter uncovers that polysemy, namely the occurrence of multiple meanings in data, arises in optimal model selection for deterministic finite automata (DFA) under MDL. This suggests that even in basic algorithmic contexts, data can be given multiple irreconcilable interpretations. This has ramifications for any applied system that can be modelled as a DFA.

**Keywords:** MDL, Grammar Induction, Polysemy, DFA, Learning

## 1 Introduction

The abstract problem of learning a grammar, or a production rule, from given data examples involves the challenge of developing algorithms that can generalise from observed data to make predictions about new, unseen instances. This process is inherently complex, and various methodologies have been explored to advance our understanding and improve the design of learning algorithms. One important method is learning by data compression [1] [30]. Essentially, this method functions as a form of pattern recognition, deriving knowledge based on the redundancy present in the data.

---

Thanks to the Univ. of Bristol Alumni Travel Grant for funding my SILFS talk.

# SMILES

Figure 1: Gestalt vase or faces. See [21].

The minimum description length (MDL) principle, in particular, is a data compression technique which seeks the model that minimises the description of the data [7]. This naturally connects the technique to principles involved in the study of algorithmic entropy (Kolmogorov complexity). The definition of algorithmic entropy involves the measurement of the information content in a data string through the identification of the minimum program size capable of generating that string when executed on a universal Turing machine [17]. It's worth noting that a universal Turing machine (UTM), a theoretical computing device with the ability to simulate any other Turing machine, is chosen arbitrarily, provided it can effectively simulate any Turing machine.

However, in the context of learning by data compression, the notion of polysemy adds a layer of complexity to the learning process. Polysemy refers to the phenomenon where data can have multiple meanings or interpretations. Such a data set would be akin to a "multistable" image as understood in the theory of Gestalt Psychology: briefly, a gestalt image is one in which there is an irresolvable conflict that defies the mind's ability to neatly work the information into a single interpretation as to what the mind is seeing. Figure 1, above, is the canonical example.

Polysemy becomes a challenge because the same data may be generalised in different, sometimes conflicting, ways. Consider a scenario where a machine learning model is trained on a dataset that exhibits polysemy. Here the algorithm needs to grapple with the fact that the description of certain data instances can be minimised in irreconcilable ways which both aggregate the same amount of compression.

This is similar to a multistable image in the following sense: there are at least two interpretations which seem to compete. There are two optimal ways to model the perceptual data in figure 1; as a black vase, or as two faces silhouetted on a black background. The data set is polysemantic. In the language of optimal model selection, this is an example of one's mind compressing the visual information into multiple coherent meanings. These connections between Gestalt psychology and data compression have been studied extensively in Adriaans [2].

The contribution of this chapter is to demonstrate that the problem of polysemy also arises in the optimal model selection problem for deterministic finite automata (DFA), which are very simple models of computation studied in theoretical computer science. DFA compute what are know as regular grammars. See the Chomsky hierarchy [9]. These are simply the easiest rules to "parse" as they involve no context-sensitivity and minimal complexity.

It would be reasonable to think such grammars would not omit polysemy because of the simplicity of their production rules. Indeed, such grammars are often involved in search algorithms or very simple mechanical computations such as the programs implemented in elevators and vending machines [11]. It seems strange to think that data sampled from either of these system could exhibit polysemy. But regular grammars also crop up in much more exotic arenas: they have been used to model protein processes [23], for instance. The question of polysemy in DFA is thus a nontrivial one.

This highlights the inherent complexity of inductive inference tasks, regardless of the underlying data production rules. The existence of multiple meanings associated with the same data sets suggests that there is no fundamental barrier in complexity preventing this issue from arising in simpler contexts. Approaches that recognise and integrate polysemy play a role in advancing the development of more flexible and adaptive learning systems.

The contents of the rest of the chapter are as follows: we first explain the history of the problem of grammar induction. Next we develop the mathematics required to understand it: the salient mathematical operation which underlies grammar induction on DFA is a partitioning of the state space corresponding to a state-merging algorithm. Moreover, it is necessary to include previous results which demonstrate the link between algorithmic complexity, learning and inductive inference. Minimising the description length of both the data, and the data when encoded with a model, involves entropic model selection. This is model selection based on algorithmic mutual information. Finally, we demonstrate that the problem of polysemy (which has previously been shown to arise in the context of unrestricted grammars [2]) also arises in the most basic grammar (regular grammars). To conclude, we offer some contextualising remarks and directions for future research.

## 2 Grammar Induction

In this chapter, we are specifically interested in the abstract learning problem known as grammar induction (for DFA). Suppose an infinite set of sequences of finite length. This could be, for instance, the set of even numbers, or the set of all finite combinations of letters from the Roman alphabet. Given some finite sampling of this set, how hard would it be, computationally speaking, to infer the structure of the infinite set? That is, how hard would it be to find a rule which produces (all, or most of) the infinitely many strings that are members of the set? If it is indeed possible to effectively learn an infinite set from a finite sample, what are the conditions that are typically required in order for learning to take place?

In the context of regular languages—viz. sets of finite strings which are modelled by finite-state automata—there is a rich history concerning the former questions: it is possible to infer a grammar for a regular language by running a state-merging algorithm which searches for the minimal consistent DFA, from a finite collection of finite strings, each labeled as to whether the DFA should accept or reject. However, an exhaustive search to find the smallest DFA consistent with an arbitrary data sample is NP-hard [5] [19], and it is NP-hard to approximate a target DFA of size OPT by a DFA of size within $OPT^k$ for any constant factor $k$ unless P=NP [20]. Phrased as a decision problem, it is NP-complete to decide, given $k \in \mathbb{N}$ and two finite sets $A$ and $B$ of words, whether there exists a DFA with at most $k$ states which accepts every string in $A$ and none of the strings in $B$ [12]. Furthermore, the cryptographic hardness of learning DFA functions from signed data examples under Valiant's [28] Probably Approximately Correct (PAC) learning model is a reduction from the problem of inverting the RSA cryptosystem, since one can encode the RSA function into a DFA [14]. This constitutes good evidence that not even the average-case DFA will be easily PAC learnable.

The evidence against PAC learning begs the question as to whether a good choice of learning model can alleviate the difficulty of learning DFA that is suggested by the worst-case analysis. Indeed, many of the positive results for learning formal languages have been achieved by additionally hypothesising constraints on the distribution of examples in the data sample, see [6] [22]. For DFA in particular, Dennis [10] showed that DFA are Probably Exactly Correct (PEC) learnable with simple positive examples. In this case, 'simplicity' is defined as having low Kolmogorov Complexity. The successful result of PEC learning implies that a heuristic based on two-part code optimisation might lead to a more successful algorithm.

The job of two-part code optimisation is to minimise the description of both the literal data and the program that is representing it. That is, to minimise the complexity of the model and the complexity of the data when predicted with the model. In both statistical and algorithmic notions of model selection, finding patterns and regularities in data is a natural part of the process of structuring data in order to fit a hypothesis or model. This is in essence how learning and data compression relate to one another. Two-part code is seemingly promising as it captures the interplay between the complexity of the model/program and the amount of compression the model offers for the literal data. However, despite the success of PEC learning, we show that, for arbitrary data, minimum description length (MDL) optimisation (which is a variant of two-part code optimisation) implies the existence of certain data sets that are not learnable for unsupervised machine learners. This is due to polysemy.

This is interesting not simply because it demonstrates an application of the computational theory of meaning captured by the expression "learning as data compression", but also because investigating regular grammars in juxtaposition to recursively-enumerable grammars informs us about the relationship between language complexity and learning: regular grammars exhibit polysemy when there is no uniquely distinguishing element in the data. By the Myhill-Nerode theorem [18], there exists a unique minimal DFA for each regular language with a number of states equal to the number of distinct equivalence classes of the language. What this suggests is that the polysemanticism we see is not attributable to the nature of regular languages in and of themselves, but rather to the nature of inferring complete structure from incomplete information, whereas one can trivially find optimal Universal Turing machines for partial and total recursive functions because of their ability to emulate other UTMs.

# 3 Mathematical Preliminaries

**Definition 1** (Finite partition) *A partition $\pi$ of a set $X$ is a set of nonempty subsets of $X$ such that every element $x \in X$ is in exactly one of these subsets. $B(x, \pi) \subseteq X$ indicates the subset of the partition $\pi$ of which $x$ is an element.*

**Definition 2** (Quotient automaton) *Let $M = (Q, \Sigma, \delta, q_0, F)$ be a deterministic finite automaton, a quotient automaton $A/\pi = (Q_\pi, \Sigma, \delta_\pi, B(q_0, \pi), F_\pi)$ derived from $M$ on the basis of a partition $\pi$ of $Q$ is defined as follows:*

(1) $Q_\pi = Q/\pi = \{B(q, \pi) \; ; \; q \in Q\}$

(2) $F_\pi = \{B \in Q_\pi \; ; \; B \cap F \neq \emptyset\}$

(3) $\delta_\pi : (Q_\pi \times \Sigma) \to 2^{Q_\pi}$ ; $\forall B, B' \in Q_\pi$, and $\forall a \in \Sigma$
$B' \in \delta_\pi(B, a) \iff \exists q, q' \in Q, q \in B, q' \in B'$ and $q' \in \delta(q, a)$.

*States of $Q$ that are partitioned into the same equivalence class $B$, are merged.*

**Lemma 3** *If an automaton $A/\pi_j$ is derived from an automaton $A/\pi_i$ by means of a partition then $L(A/\pi_i) \subseteq L(A/\pi_j)$.*

**Definition 4** (State-merging DFA induction algorithm) *A State-merging DFA induction algorithm starts from an initial automaton referred to as a prefix tree acceptor $PTA(D_+)$. The PTA is the largest trimmed DFA accepting exactly the positive data examples, $D_+$. The algorithm then merges states in the PTA relative to a partition $\pi$ of the state set of the original automaton. States belonging to the same block of $\pi$ are merged in the resulting quotient automaton. Any accepting path in the PTA is also an accepting path in $PTA/\pi$. This clear definition is found in* [15].

**Definition 5** (The Minimum Descriptive Length Principle) *The best theory to explain the data is the one which minimises the sum of the length in bits of the description of the theory, and the length in bits of the data when encoded according to the theory* [4].

Note this is the two-part code definition of the MDL principle as it occurs in statistical learning and machine learning generally. We now explain more fully how the principle relates to algorithmic information theory and inductive inference. See [3] for an in depth discussion of the principle's strength and weaknesses.

Let $M \in \mathcal{M}$ be a model in a class of models, and let $D$ be a data set. The prior probability of a model or hypothesis is $P(M)$. This is the likelihood we ascribe to the model before any observations have been made. $P(D)$ is the probability of the data, defined with respect to all models which are consistent with the data. The posterior probability is $P(M|D)$, that is the probability we ascribe to the model given the data we have observed. The maximum *a posteriori* probability is the $argmax$ function of this probability.

$$M_{MAP} \equiv argmax_{M \in \mathcal{M}} P(M|D) \tag{1}$$

$$= argmax_{M \in \mathcal{M}} \left( \frac{P(D|M) \cdot P(M)}{P(D)} \right) \tag{2}$$

$$\equiv argmin_{M \in \mathcal{M}} \left( -\log P(M) - \log P(D|M) \right) \tag{3}$$

(2) Follows by Bayes' Law. Since $D$ is constant over $M \in \mathcal{M}$ and log is always increasing, we can take the logarithm of argmax without loss of generality. In terms of Shannon code, the derived MAP says that the minimal (optimal) amount of bits required to encode the model is $-\log P(M)$, and the optimal length in bits to encode the model given the data is $-\log P(D|M)$. We call the former, the *model code*, and the latter the *data-to-model code*. The MAP we have derived to be the minimum $M \in \mathcal{M}$ of the summed lengths of the model and data-to-model code. But this is exactly the definition of MDL. Hence $M_{MDL} \equiv M_{MAP}$.

Of course, the argmax equation is reliant on the class of models, which we have yet to specify. If we were to specify $\mathcal{M}$ as consisting of an enumeration of all self-delimiting programmes for a preselected arbitrary universal Turing machine $\mathcal{U}$, then we would be dealing with one of the most general classes. Under this interpretation of $\mathcal{M}$, the length of the optimal two-part code is given by the Kolmogorov complexity [4].

**Definition 6** (Conditional Kolmogorov Complexity) *The* conditional prefix Kolmogorov complexity *of $x$ given $y$ is*

$$K(x|y) = min_p \{ \ell(p) \mid \mathcal{U}(\langle p, y \rangle) = x, p \in \{0,1\}^* \}$$

*Where $\mathcal{U}$ is a reference universal prefix-free Turing machine, and $\ell(p)$ is the length (in bits) of program $p$ running on $\mathcal{U}$. Note: by this definition, we can also understand the non-conditional prefix Kolmogorov complexity to be $K(x) = K(x|\epsilon)$ since $\mathcal{U}(\langle p, \epsilon \rangle) = \mathcal{U}(p)$, that is to say that it is equal to the conditional Kolmogorov complexity with empty input.*

We know from [16] that Solomonoff's universal probability, outlined in [25], is related to Kolmogorov complexity by the equation $-\log m(x) = K(x) + O(1)$. Thus replacing the information-theoretic description of Bayes' Law for an unspecified probability measure, with Solomonoff's universal probability yields the following equation for the algorithmic probability version of Bayes' Law:

$$-\log m(x) + \log m(x|y) = K(x) - K(x|y) + O(1).$$

This is the same as the definition of mutual algorithmic information (within constant time complexity, $O(1)$).

**Definition 7** (Kolmogorov's Mutual Algorithmic Information) *The information in y about x is defined as*

$$I(y:x) = K(x) - K(x|y^*)$$

*where $y^*$ denotes the first (in a standard enumeration order) shortest prefix programme that generates y and then halts* [8] [13].

Thus, we have arrived at the result that Bayes' Law is actually a very foundational notion indeed; it is equivalent to Kolmogorov's algorithmic mutual information (see [8] for further discussion). To understand this in less formal terms: inductive inference can be understood as finding the model that maximises the *a posteriori* probability conditioned on randomly observed data. This is indeed what is captured by Bayes' law; the more correlated the data are with the chosen model, then the more certain is your posterior hypothesis. Bayes' law in one sense is simply a way to update one's beliefs in accordance with observation. That in itself is a form of learning.

The connection between finding the best model which fits the data, as well as updating one's beliefs in accordance with observation is what relates the maximum *a posteriori* probability to Kolmogorov complexity. This is because Kolmogorov complexity is itself defined with respect to a minimal program that compresses the data. This implies that Kolmogorov compelxity provides an axiomatic foundation of inductive inference, since it is, in essence, a formalisation of Occam's razor.

It is also fundamental to the learning by data compression for the reason that Kolmogorov complexity is defined as the maximum compression avaliable in a string of data. Since Bayes' Law gives a kind of algorithm for updating one's beliefs, looking for models which maximise the posterior probability will inevitably relate algorithmic notions of learning with algorithmic data compression.

For grammar induction on DFA, the general idea is to measure the descriptive complexity of an automaton along with the descriptive complexity of encoding a set of sample words using the automaton as a model. It will not always be the case that the smallest automaton is recognised as the most promising, since the complexity of the process of parsing strings in the sample is also quantified.

In our context, we seek to deal with finite automata. Hence $M \in \mathcal{M}$ will specify an automaton in the class of all deterministic finite automata. From now on, we will interpret $\mathcal{M}$ in this sense. Seeing as, in this context, programmes are in general not self-delimiting, it will not be appropriate to concatenate programmes at will. Here the Shannon code length is the most appropriate measure of MDL. This is simply because the exact Kolmogorov complexity is uncomputable; it is defined as the shortest program the produces the string when input on some reference universal Turing machine. Knowing which programs are the shortest would require knowledge of the halting programs, which of course is ruled out by Turing uncomputability [27]. However, the expected value of Kolmogorov complexity equals its Shannon entropy up to constant [26]. Thus, the former can be approximated by the latter.

# 4 Polysemy in DFA

Following [29], let $A = (Q, \Sigma, \delta, q_0, F)$ be a DFA all of whose non-final states have outgoing edges. The number of bits required to encode the path traversed in order to parse a word $w$ can be assessed by the function $ch$ given below. We associate with each state $q \in Q$ the value $t_q = \sum_{a \in \Sigma} |\delta(q,a)|$ if $q \notin F$. Whereas, if $q \in F$ then $t_q = 1 + \sum_{a \in \Sigma} |\delta(q,a)|$, since one more choice is available, namely the choice to accept. We are now in a position to define $ch(q,w)$. For the empty word we have $ch(q,\epsilon) = \log(t_q)$ if $q \in F$; otherwise $ch(q,\epsilon) = \infty$. For $w = au (a \in \Sigma, u \in \Sigma^*)$, the function depends on the recursive definition: $ch(q,w) = \log(t_q) + min_{r \in \delta(q,a)} ch(r,u)$ and $c(q,w) = \infty$ if $\delta(q,a) = \emptyset$. Given a sample $D_+$ and a DFA $A$, we can now measure the MDL score $sc$ of $A$:

$$sc(A, D_+) = |Q| + ||\delta||(2\log|Q| + \log|\Sigma|) + \sum_{w \in D_+} ch(q_0, w)$$

where $||\delta||$ is the number of transitions of $A$. Note, see Wieczorek [29] for details on this choice of MDL two-part code optimisation for DFA induction.

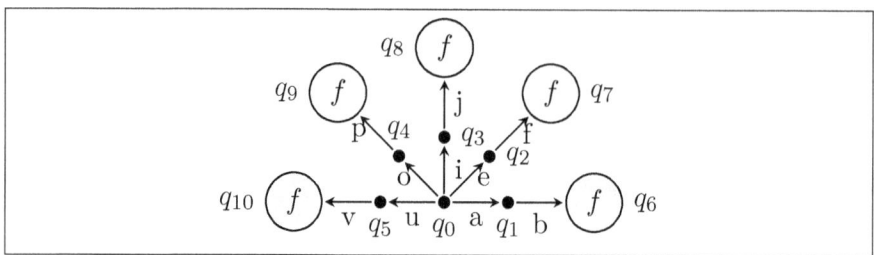

Figure 2: The PTA of $D_+^0$

**Definition 8** (Polysemantic data set) *A data set $D$ is polysemantic if there exist multiple optimal codifications $M_i \in \mathcal{M}$ for $D$ such that the mutual information between any two $M_i$ is small, or zero, i.e. $I(M_1 : M_2)$ is near to or equal to $D$.*

**Theorem 9** *Regular languages are polysemantic under MDL.*

In principle, just one counterexample is sufficient to disprove the claim that regular languages are *not* polysemantic under MDL. This would then imply, under the assumption that languages are regular, that there are cases where regular languages are also polysemantic.

*Proof.* Suppose we have a data set made up from positive *and* negative examples (in order to control for overgeneralisation). Let us define a data set $D_+^0 = \{ab_+, ef_+, ij_+, op_+, uv_+\}$ and $D_-^0 = \{b_-, f_-, j_-, p_-, v_-\}$. With $D^0 = D_+^0 \cup D_-^0$. The PTA can be formed as usual: see figure 2 above.

The data seem to have the pattern "vowel followed by consonant", and the rule "never consonant without a vowel prefix." There are two natural choices for regular languages which might describe this: $L_1 = (Vowel) \circ (Consonant)^*$, or, $L_2 = [(Vowel) \circ (Consonant)]^*$. DFA models for these languages can be found by merging states.

The merge of $q_i \in B(q_f, \pi)$ $\forall i : 0 < i \leq 10$, and $q_0 \in B(q_0, \pi)$ yields the DFA representing $L_1$ (figure 3). The merge of $q_0, q_n \in B(q'_f, \pi')$ $\forall n : 5 < n \leq 10$, and $q_n \in B(q', \pi')\forall n : 0 < n \leq 5$ yields the DFA representing $L_2$ (figure 4).

We now have two, 2-state DFA which are consistent with $D^0$. Note that

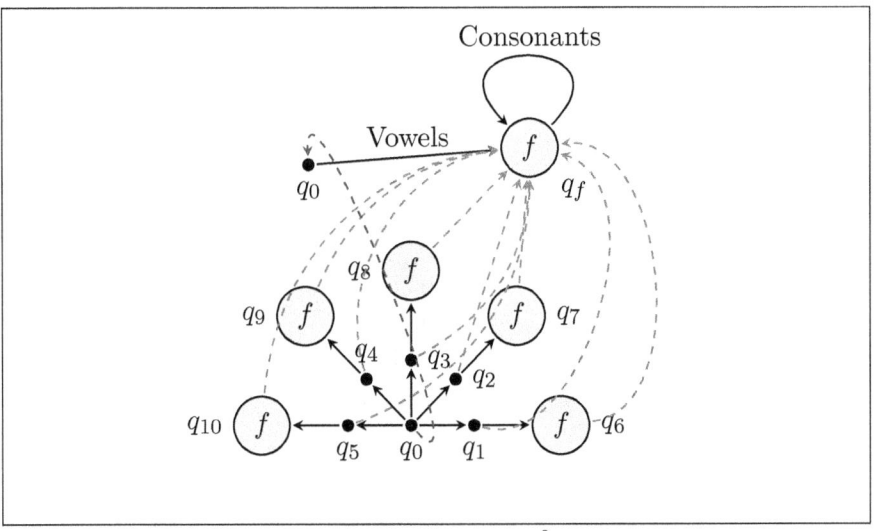

Figure 3: The merge of $D_+^0$ into $L_1$

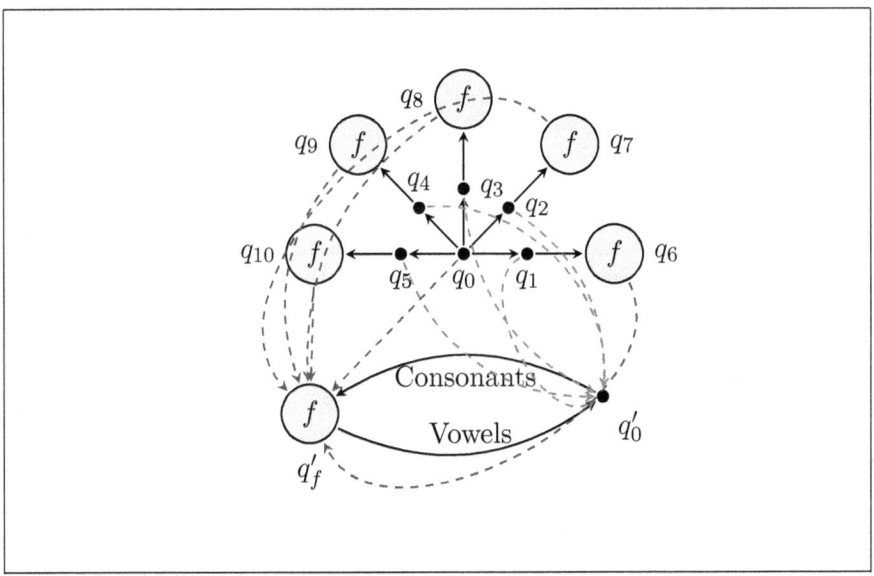

Figure 4: The merge of $D_+^0$ into $L_2$

$L(A_1) \cap L(A_2) = D^0$, there are no other words in the intersection. Indeed, in $A_1$ a consonant can follow a consonant which is not possible in $A_2$, whereas in $A_2$ a vowel can follow a consonant, which is not possible in $A_1$. We now show that $A_1$ and $A_2$ are equally parsimonious optimal descriptions of $D^0$ despite being disjoint.

We have $t_{q'_f} = t_{q_f} = 6$ and $t_{q_0} = t_{q'_0} = 5$. However, since $q'_f$ and $q_f$ are the final states of $L_2$ and $L_1$ respectively, and likewise $q'_0$ and $q_0$ are the non-final states, it is clear that the cost of encoding $D_+$ on either $A_1$ or $A_2$ will be the same, as encoding $D_+$ according to either model nullifies what might be saved by encoding in the other model. Since the number of states, transitions and alphabet are also identical for $A_1$ and $A_2$, we can be sure $sc(A_1, D_+) = sc(A_2, D_+)$. This is, incidentally, 93 bits rounded to the nearest whole bit.

Furthermore, we can be sure that there can be no smaller model, since the universal automaton is inconsistent with $D^0_-$. There is a 3-state automaton generating a *finite* language equivalent to the PTA which has slightly lower descriptive complexity (91 bits). However, this finite language can be ruled out as it offers no generalisation. Thus $D^0$ is polysemantic under MDL.

□

## 5 Conclusions

We are able to conclude that regular languages are polysemantic under two-part code optimisation given by MDL. Putting it all together, if regular languages are considered polysemantic under MDL, it means that when using MDL to learn or describe regular languages, the learning process must account for the fact that certain data instances may have multiple valid interpretations. The MDL principle, in seeking simplicity and accuracy in the model and data-to-model code, must take into consideration the inherent complexity introduced by the possibility of multiple meanings in the data. This recognition of polysemy is crucial for developing learning algorithms that can effectively handle the inherent intricacies of inference problems even in the very simple context of regular languages.

Beyond this, the presented work in this chapter provide one with a general picture of the role of algorithmic entropy (Kolmogorov complexity) in optimal model selection which transcends the specific learning problem we have answered: irrespective of the simplicity of the model and the model selection problem, there seems to be a general relationship of data, sample and population which is not based strictly in statistics or probability, but in the deeper framework of algorithmic learning. Specifically, learning with insufficient information leads to situations where entropy becomes invariant over optimal models. In such cases there are polysemantic interpretations to the data.

Further investigation into formal language theory could question whether are there any grammars which can be learned using entropic model selection techniques where polysemy cannot arise. Perhaps in the context of subgrammars of regular grammar, such as star-free grammar, it is simply not possible to interpret the data in multiple ways.

Nonetheless, it is already a little surprising that data for DFA exhibit polysemy, as such machines follow very simple rules and only have finite memory. The interesting thing to extrapolate is that any process that can be modeled as a DFA has the potential to churn out data sets that are polysemantic. One modern example might be the use of DFA modelling of protein processes [24]. If an algorithm were created that attempted to identify a protein process as a DFA, and the given data are polysemantic, in this case there will be multiple DFA that may share little to no mutual information that are all optimal models for the data. At first this seems to imply a problem in modelling such processes as DFA. However, another interesting avenue could be to research whether biomachinery can trade on the fact that certain data can be interpreted variously; namely, if one signal can take on multiple meanings depending on other ad hoc contextual properties. In this case it would seem efficient of nature to utilise the ambiguity within signalling to its advantage.

Pattern recognition and model selection exhibit polysemanticism because they involve perspective, but this does not make the phenomenon purely epistemological; apprehending the world is itself a form of data compression, and necessarily requires both the representer and the represented. This is to say that representation is *necessarily* perspectival. For that reason polysemy is partly ontic too. This may mean that the 'true' nature of certain phenomena cannot be found in the data themselves no matter the size and quality of the sample collected, which is itself something to consider when pondering the success of data-driven learning models which aim to understand complex phenomena such as language use or signalling more generally.

# 6 Bibliography

[1] P. W. Adriaans. Learning as data compression. In S. Barry Cooper, Benedikt Löwe, and Andrea Sorbi, editors, *Computation and Logic in the Real World*, pages 11–24, Berlin, Heidelberg, 2007. Springer Berlin Heidelberg.

[2] P. W. Adriaans. A computational theory of meaning. In *Advances in Info-Metrics: Information and Information Processing across Disciplines*. Oxford University Press, 2020.

[3] Pieter Adriaans and Paul Vitanyi. The power and perils of mdl. In *2007 IEEE International Symposium on Information Theory*, pages 2216–2220, 2007.

[4] Pieter W. Adriaans and Ceriel J. H. Jacobs. Using mdl for grammar induction. In *International Conference on Graphics and Interaction*, 2006.

[5] Dana Angluin. On the complexity of minimum inference of regular sets. *Information and Control*, 39:337–350, 1978.

[6] Dana Angluin. Inference of reversible languages. *Journal of Applied and Computational Mechanics*, 29:741–765, 1982.

[7] A. R. Barron, J. Rissanen, and B. Yu. The minimum description length principle in coding and modeling. *IEEE Transactions on Information Theory*, 44:2743–2760, 1998.

[8] Fouad B. Chedid. Kolmogorov's algorithmic mutual information is equivalent to bayes' law. *ArXiv*, abs/1907.02943, 2019.

[9] N. Chomsky. Three models for the description of language. *IRE Transactions on Information Theory*, 2:113–124, 1956.

[10] François Denis. Learning regular languages from simple positive examples. *Machine Learning*, 44:37–66, 2001.

[11] Preye Ejendibia and Barilee Baridam. String searching with dfa-based algorithm. *International Journal of Applied Information Systems*, 9:1–6, 10 2015.

[12] E. Mark Gold. Complexity of automaton identification from given data. *Information and Control*, 37:302–320, 1978.

[13] Peter D. Grünwald and Paul M. B. Vitányi. Shannon information and kolmogorov complexity. *ArXiv*, cs.IT/0410002, 2004.

[14] Michael Kearns and Leslie G. Valiant. Cryptographic limitations on learning boolean formulae and finite automata. In *Journal of Applied and Computational Mechanics*, 1994.

[15] Bernard Lambeau, Christophe Damas, and Pierre Dupont. State-merging dfa induction algorithms with mandatory merge constraints. In *International Conference on Graphics and Interaction*, 2008.

[16] L. A. Levin. Laws of information conservation (nongrowth) and aspects of the foundation of probability theory. In *Problems of Information Transmission*, pages 206–210, 1974.

[17] M. Li and P. Vitányi. *An Introduction to Kolmogorov Complexity and Its Applications*. 01 1997.

[18] Anil Nerode. Linear automaton transformations. In *Proceedings of the American Mathematical Society*, volume 9, pages 541–544, 1958.

[19] Leonard Pitt. Inductive inference, dfas, and computational complexity. In *Analogical and Inductive Inference*, 1989.

[20] Leonard Pitt and Manfred K. Warmuth. The minimum consistent dfa problem cannot be approximated within any polynomial. *[1989] Proceedings. Structure in Complexity Theory Fourth Annual Conference*, pages 230–, 1989.

[21] E. Rubin. *Synsoplevede Figurer*. 1915.

[22] Yasubumi Sakakibara. Efficient learning of context-free grammars from positive structural examples. *Information and Computation*, 97:23–60, 1992.

[23] W. Samarrai, J.W. Yeol, I. Barjis, and Y.S. Ryu. System biology modeling of protein process using deterministic finite automata (dfa). In *2005 9th International Workshop on Cellular Neural Networks and Their Applications*, pages 290–295, 2005.

[24] Wallied Samarrai, Joe Woong Yeol, I. Bajis, and Yeong S. Ryu. System biology modeling of protein process using deterministic finite automata (dfa). *2005 9th International Workshop on Cellular Neural Networks and Their Applications*, pages 290–295, 2005.

[25] Ray J. Solomonoff. A formal theory of inductive inference. part ii. *Information and Control*, 7:224–254, 1964.

[26] Andreia Teixeira, Armando B. Matos, André Souto, and Luis Filipe Coelho Antunes. Entropy measures vs. kolmogorov complexity. *Entropy*, 13:595–611, 2011.

[27] A. Turing. On computable numbers, with an application to the entscheidungsproblem. *The Proceedings of the London Mathematical Society*, s2-42:230–265, 1937.

[28] Leslie G. Valiant, David Pollard, and Adolfo J. Quiroz. A theory of the learnable. In *Communications of the American Mathematical Society*, volume 11, page 1134–1142, 1972.

[29] Jan van Leeuwen and Arlindo L. Oliveira. Grammatical inference: Algorithms and applications. In *Lecture Notes in Computer Science*, 2000.

[30] J. Wolff. Unifying computing and cognition: The sp theory and its applications. *ArXiv*, cs.AI/0401009, 2004.

# FEYNMAN'S THEORIZING AND VISUALIZATION

MARCO FORGIONE
*University of Milan*
marco.forgione@unimi.it

**Abstract.** In this paper I explore and discuss the type of scientific understanding that Feynman had when he was developing his famous diagrams. It is argued that such understanding came from the physicist's capacity of visualizing the phenomena and that such visualization-skill contributed to the forming of a narrative explanation. In support of this view, I consider two historical examples: (1) the absorber theory of radiation and path integrals, (2) and the lackluster reception of Feynman's diagrammatic method at the Pocono conference (1948).

**Keywords:** Feynman Diagrams, Scientific Representation, Visualization and Understanding.

## 1 Introduction

With the present contribution I wish to explore the understanding of quantum phenomena that Feynman had when he was developing his famous diagrams [2], which are graphical representations of the transition amplitude of an interaction between quantum (subatomic) particles.[1] I will argue that such understanding came from the physicist's capacity of visualizing the phenomena and that such visualization-skill contributed to the forming of a narrative explanation in the sense of [3] and [4].

---

This paper is part of a project that has received funding from the European Research Council (ERC) under the Horizon 2020 research and innovation programme (Grant Agreement No. 758145)

[1] More accurately, Feynman diagrams represent the perturbative terms used to calculate the transition amplitude.

Notably, the importance of visualization in physical theories has already been acknowledged and discussed in the literature. For example, Boltzmann [1] believed that theories should provide pictures of the physical world that guide scientific thinking and experiments. These pictures, though, should not be considered as faithful representations (one-to-one) of physical phenomena. Similarly, Schrödinger believed that spacetime visualizability contributes to the making of good scientific theories —even though the theories are not representative of the real world. Yet, how do these considerations apply to the case of Feynman diagrams?

Feynman's theorizing was such that the explanation of a given phenomenon, that is, the writing of mathematical equations attuned with the empirical data, came from a form of visualization of phenomena, and thus from a partial representation of physical reality. Such an understanding was originally intended as providing an answer to plausible explanatory why-questions in the form of stories [4] that link models (and theories) to physical objects (and phenomena).

In support of this reading, I will consider some historical examples: (1) Feynman's attempt to understand quantum phenomena was based on his theories on classical electrodynamics, path integrals, and the quivering electron. (2) Feynman's presentation at the Pocono conference was ill-received because it made use of the diagrammatic simplifications to avoid some mathematical complexities and the audience was not used to such visual-based thinking. The lackluster reception of his talk forced Feynman to publish his works in a mathematically more rigorous way. What will emerge from examples (1) and (2) is that the use of visualization techniques was fundamental to Feynman's scientific method and theory building.

## 2 Feynman's Take on the Phenomena

Before Feynman diagrams, the calculation of new terms in a perturbative series for the transition amplitude of a scattering event could constitute a significant research project. Indeed, it could even be a substantial part of a doctoral dissertation in theoretical physics. Physicists worked on extending the perturbation series to higher orders, improving the precision of predictions, and developing more sophisticated mathematical techniques for handling complex interactions. This work was valuable for advancing the field, but it often required a deep understanding of the formalism of quantum field theory, as well as advanced mathematical methods. For example, the formalism of quantum field theory involved the manipulation of creation and annihilation operators as well as solving integrals over spacetime. Such algebraic manipulations required abstract mathematical concepts, while the number of terms in a perturbative series could grow rapidly when dealing with multiple particles.

The use of Feynman diagrams significantly facilitated the calculation of transition amplitudes by providing a visual and systematic way to represent and organize the terms in the perturbative expansions. For example, the fact that each vertex in a diagram corresponds to an interaction term (and each line to a propagator) allowed physicists to keep track of the different mathematical expressions, resulting in a systematic organization of the calculations. In addition, the diagrams naturally respect general principles such as conservation of momentum, charge, and energy at each vertex. This ensures that the mathematical expressions derived from the diagrams also respect the same principles. In general, Feynman's introduction of the diagrams provided a revolutionary new way to organize and visualize the calculation of transition amplitudes. While they allowed physicists to systematically represent and calculate terms in the perturbation series, making the process more intuitive and accessible, their graphical nature facilitated communication and collaboration among researchers.

However, the initial reaction to Feynman's new method was at best cold. Perhaps, this was made most evident at the Pocono conference in Pennsylvania (1948) when Feynman presented a new approach to quantum electrodynamics to his peers.

> At each step he was asked to justify his procedure; instead he offered to work out a physical example to demonstrate the correct results it produced [...] The culmination of his audience's feeling that Feynman was running amok without being rigorous came when Niels Bohr stood up, objected to Feynman's use of trajectories for small particles, and started reminding him about Heisenberg's uncertainty principle. Here Feynman gave up in despair, realizing that he couldn't communicate the fact that his analysis was justified by its correct results [5, 495].

In spite of the initial rejection, Feynman diagrams started to spread out among scientists, especially thanks to the work by Dyson [6], who proved the equivalence between the method suggested by Feynman and the theory of Schwinger and Tomonaga. Dyson also formalized the diagrams in the form of formal rules, a step-by-step guide for calculating scattering amplitudes [7]. As remarked by Kaiser [8, 76]:

> [...] Dyson's pair of articles became a "how-to" guide, enumerating carefully the step-by-step rules for drawing Feynman's new diagrams and explaining how their bare lines were to be translated uniquely into the mathematical elements of scattering amplitudes. It was thus Dyson, and not Feynman, who first codified the rules for the diagrams' use —precisely what Feynman's frustrated audience had hoped to hear at the Pocono meeting a few months earlier.

There is, though, an important difference between the significance bestowed to the diagrams by Feynman and by Dyson. On the one hand, Feynman favored a physical interpretation of the diagrams and initially interpreted them as depicting physical processes. On the other hand, Dyson interpreted the diagrams as a set of calculation rules and used them as graphical representations of combinatorial possibilities. It is again Kaiser [8] that remarks the history of the split between the two physicists:

> From the very beginning, Feynman and Dyson held different ideas about how the diagrams should be drawn, interpreted, and used [...] To Feynman, his new diagrams provided pictures of actual physical processes, and hence added an intuitive dimension beyond furnishing a simple mnemonic calculational device. To Dyson, the line drawings were never more than "graphs on paper" handy for manipulating long strings of equations but not to be confused with the stuff of the real world [8, 176].

In sum, the lackluster reception of the diagrams testifies Feynman's original method of using graphical representations for phenomena that were described entirely mathematically. This new method was ill-received by the scientific community since the fundamental principles of the theory, such as the uncertainty principle, were an obstacle to any graphical representation. In addition, the divide between Dyson and Feynman suggests that the latter interpreted the graphical representations as indicating the presence of a specific quantum phenomenon, and not as a mere calculation tool. This is not to say that Feynman was unaware of the uncertainty principle, but rather that he thought of his diagrams as partial representations of scattering events. In the next section, I will maintain that Feynman's original method of modeling phenomena can be traced back to his previous works on classical electrodynamics and quantum mechanics. But why is the focus on phenomena relevant for characterizing Feynman's theorizing? By focusing on phenomena, Feynman was forced to imagine and visualize what he thought was happening at the physical level. In this sense, the equations represented a translation of what he though was 'physically happening' into a formal language. This is in contrast to the more common attitude of exploring the mathematical consequences of a given theory and on the construction of new mathematical tools to accommodate the empirical data.

## 2.1 Absorber Theory and Quivering Electron

As a first example, we should consider the development of Feynman's absorber theory of radiation in [10] and [11]. The motivation for developing a new theory of classical electrodynamics was to solve the problem of the self-action of the electron with its own filed, and the consequent divergence of the field strength: "it seemed to me quite evident that the idea that a particle acts on itself, that the electrical force acts on the same particle that generates it, is not necessary one —it is sort of a silly one, as a matter of fact" [12, 2]. The rejection of the self-action led Feynman to develop a theory based on the direct-action-at-a-distance between individual particles, and on the retarded and advanced solution to Maxwell equations.[2] The gist of the theory is that an accelerated charged particle emits an outgoing radiation that moves forward-in-time. The radiation is then received by absorbers surrounding the source and re-emitted as advanced and retarded radiation back to the initial source. The advanced radiation will reach the source at the time of the initial emission, thereby constituting the the so-called radiative reaction.

With the absorber theory of radiation, Feynman obtains a new description of radiative phenomena, and the rejection of a narrative that explains the radiative reaction as a force exerted by the field generated by the accelerated source. Conversely, the narrative conveyed by Feynman's theory explains the radiative damping in terms of the advanced response of the absorbers surrounding the source to the initial radiation of the accelerated charged particle. The term 'narrative' refers to that part of a model that conveys what is represented by that model (or theory). In the specific case of the absorber theory of radiation, the narrative explains the radiative phenomena by telling a phenomenological story about action-at-a-distance and the exchange of retarded-advanced radiation between particles. The story conveyed by the narrative is warranted by the mathematical structure of the theory (or model) and, as such, it provides a physical interpretation of the equations of the same theory (or model).

---

[2] A detailed commentary on the derivation of the absorber theory of radiation can be found in, for example: [13] and [14].

Before moving on, I should mention that the concept of narrative and phenomenological story is introduced in, among others, [4], [3], and [15]. For example, Morgan [4, 361] discusses the explanatory role of stories in economic models: "[...] the way models help us to understand the economic world in which we live in is by telling stories about the world. That story might be a story about the real world (past, present and future), or it may be a story about the hypothetical world portrayed in the model: the relationship of the story to the model structure is the same". When applied to the mathematical formalism, the use of narratives can convey a physical interpretation to the equations, and physical interpretations are desiderata that Feynman actively sought in his theorizing. This leads us to Feynman's next theory, and to our next example.

Feynman [16] condenses and presents the main results of the dissertation; the last section of his paper offers a tentative generalization of the new path integral method to relativistic moving particles.[3] Indeed, while Feynman tries to derive an action functional for the relativistic Dirac equation, he remains dissatisfied with the impossibility of giving a physical interpretation to such derivations.

> These results for spin and relativity are purely formal and add nothing to the understanding of these equations. There are other ways of obtaining the Dirac equation which offer some promises of giving a clearer physical interpretation to that important and beautiful equation [16, 387].

Wüthrich emphasizes that the attitude displayed by Feynman is "one that is not satisfied by formal solutions but rather was after something like a model or mechanism of the processes that the formal apparatus was supposed to describe" [17, 507-508]. In addition, Feynman will later formulate a physical model of the electron to derive an action functional to the Dirac equation: the so-called quivering electron. This an attempt to understand and obtain the Dirac equation from an initially one-dimensional model of an electron moving either to the left or to the right on a lattice.[4] Again, Feynman's attention was directed towards finding a physical model that could satisfy the Dirac equation.[5]

---

[3]The path integral is a formulation of quantum mechanics that Feynman obtained from his failed attempt to quantize the absorber theory of radiation.

[4]Notably, that an electron as described by the Dirac equation oscillates around a mean trajectory was already suggested in [19] and [18]. Also, a thorough analysis of Feynman's quivering electron can be found in: [20].

[5]This is analogous to how the absorber theory of radiation offered a new ex-

In addition, Stöltzner [21] emphasizes how Feynman [16] already realized that the trajectories involved in the path integrals are continuous but not differentiable, and thus they could be interpreted as a quantum Brownian motion.[6] In this sense, the formulation of a mechanical model of the phenomena (the quivering electron) came with a form of understanding of that phenomena (quantum Brownian motion), which then led to either a formulation of a new theory, or to fitting the model within a more general theoretical framework. As argued by Wütrich [20, 65]: "[...] what Feynman means by "understanding" the Dirac equation is not the study of the mathematical properties of the Dirac equation or the search for an ingenious method of solution required by the application of the equation to complex problems. Rather, Feynman is looking for a physical system, the appropriate description of which would satisfy the Dirac equation".

Similarly to what happened at the Pocono Conference, it is uncharitable to think of Feynman as believing that the zigzagging of electrons to the right and to the left on a lattice is a truthful representation of how electrons propagate. Therefore, we can characterize the quivering electron and the comparison with Brownian motion as an attempt to formulate a phenomenological story that provided Feynman with a physical interpretation of the corresponding equations. In these terms, the physical interpretation, conveyed by the narrative, constitutes a partial representation of the phenomena.

In the next section, I will move back to Feynman diagrams and I will discuss in what sense they constitute a narrative and a partial representation of scattering phenomena.

## 2.2 Feynman Diagrams

Most recently, de Regt [23, 251-252] maintains that Feynman diagrams provide "a visualization of interaction processes, albeit one that cannot be taken as one-to-one representation of actual occurrences in nature. [...] The visual Feynman diagrams functioned as conceptual tools that made quantum field theory more intelligible for most theoretical physicists. [...] Rather than realistic representations of physical processes, Feynman diagrams are tools for solving problems and making calculations". The quote suggests that diagrams do not have a relation with physical reality and that they are not

---

planatory narrative to radiative phenomena, though, still satisfying the equations of classical electrodynamics.

[6] A deeper analysis of the mathematical character of path integrals and the non-differentiable trajectories can be found in, for example: [22].

representations (in any sense) of the physical processes they depict. In this sense, we should interpret the diagrams as tools to calculate the scattering amplitude of some phenomena. However, that the diagrams are not a one-to-one representation does not rule out weaker forms of representation of an interaction process.

This debate calls back Boltzmann's *Bildtheorie*, which maintains that there is no identity relation between the picture of the scientific theory and reality (see: [1], [24] and [25]). Generally speaking, Boltzmann suggests that we should find pictures that represent phenomena as accurately as possible, and not the absolute truthful theory. With this, he remarks his difference from a naive realist conception of scientific theories and thus rejects the one-to-one correspondence between theory and physical reality.

> Task of theory consists in constructing a picture of the external world that exists purely internally and must be our guiding star in all thought and experiment [1, 33]. No theory can be objective, actually coinciding with nature, but rather, [...] each theory is only a mental picture of phenomena, related to them as sign is to designatum [1, 90-91].

Consider, for example, the diagram of the positron-electron scattering in Figure 1. On the one hand we can easily argue that the diagram cannot be a faithful representation of an actual scattering process —for example by recalling the argument that Bohr raised against Feynman at the Pocono conference. While the diagrams depict well-defined trajectories for the quantum objects, those trajectories are not possible in quantum mechanics due to Heisenberg's uncertainty principle. [7] On the other hand, we can read the diagram as representing the annihilation of an electron-positron pair (time flows from left to right), the formation of a virtual photon, and the decay of the photon into another positron-electron pair.

While the diagrams are not a one-to-one representation of nature, the narrative they convey allows us to infer some qualitative consequences with respect to the quantum event. For example, a more precise calculation of the scattering amplitude requires adding subsequent orders to the perturbative series. This can be understood as adding a spontaneous emission and absorption of a virtual photon along the trajectory of one of the electrons.

---

[7]The same argument is also mentioned by Brown [26, 430]: "In quantum mechanics (as normally understood) the Heisenberg uncertainty relations imply that no particle could have a position and a momentum simultaneously, which means there are no trajectories, paths through space-time".

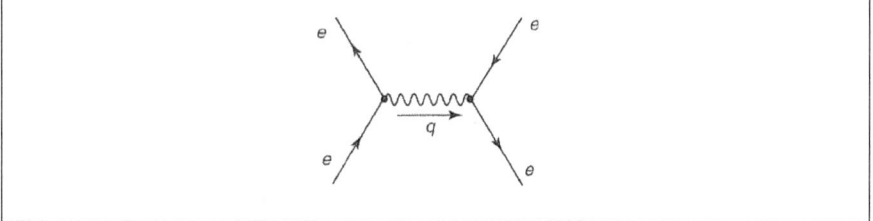

Figure 1: Second order positron-electron scattering.

In [5, 465], Feynman well summarizes the use of his diagrams and the role they play in his theorizing:

> It is hard to believe it, but I see these things not as mathematical expressions but a mixture of a mathematical expression wrapped into and around, in a vague way, around the object. [...] In certain particular problems that I have done it was necessary to continue the development of the picture as the method before the mathematics could really be found.

The fact that Feynman prioritized the picture method —the partial representation of phenomena— over the rigorous mathematical calculations is characteristic of his theorizing. While the audience at the Pocono conference expected mathematical rigor before physical interpretation, Feynman presented his results the other way around. This also explains the relevance of Dyson's work: it provided a translation of Feynman's understanding of the phenomena into rigorous mathematical equations (and proper how-to-instructions).

Before concluding, I should say a few words about whether Feynman maintained the belief that his diagrams were a partial representation of physical phenomena. The answer, I believe, is in the negative, for the physicist seemed to move toward an instrumentalist view.

## 3 A Change of Heart

There is room for arguing that the extent to which Feynman believed in the representational character of his theories and their philosophical underpinning has changed over time. As a matter of fact, if one is to accept the argument I have presented in this contribution, it should be clear that Feynman did not consider his diagrams as mere calculation tools for keeping track of the perturbative terms. Dyson, on the other hand, took the diagrams to be graphical representations of combinatorial possibilities, i.e., useful tools to manipulate the mathematical terms of the perturbative series.

There is historical evidence that shows how Feynman changed his mind about the representational character of his theories. For example, Schweber [5, 503] reports a letter written by Feynman to Wheeler in which the former rejects one of the assumptions of the absorber theory of radiation:

> I wanted to know what your opinion was about our old theory of action at a distance. It was based on two assumptions:
> - (1) Electrons act only on other electrons
> - (2) They do so with the mean of retarded and advanced potentials
>
> The second proposition may be correct but I wish to deny the correctness of the first.

Schweber [5] describes the letter as the evidence that a chapter in Feynman's intellectual life had come to a conclusion. While I agree with Schweber, I also believe that the letter conveys more than just that. First, it testifies that Feynman initially believed in the assumptions of the absorber theory of radiation. Second, it shows that Feynman had progressively questioned the extent to which his theories represented physical reality.

The second point is further emphasized when Feynman recollects his calculation efforts about the Lamb shift experiment. There, he clearly characterizes his theory as a calculation tool: "The rest of my work was simply to improve the techniques then available for calculations, making diagrams to help analyze perturbation theory quicker" [12, 14]. Similarly, the Nobel lecture [12, 16] also recalls the work done on the calculation of the interaction of the electron with the neutron:

> That was a thrilling moment for me, like receiving the Nobel Prize, because that convinced me, at last, I did have some kind of method and technique and understood how to do something that other people did not know how to do. That was my moment of triumph in which I realized I really had succeeded in working out something worthwhile. [12, 16]

The quote remarks how Feynman, in the end, and after having formulated his version of quantum electrodynamics, came to the conclusion that what he had developed was a calculation method rather than a new theory. Another remark that seems to confirm this conclusion is quoted in [9, 453] wherein

Feynman expresses his enthusiasm about his results on weak interactions in 1957.[8]

> As I thought about it, as I beheld it in my mind's eye, the goddamn thing was sparkling, it was shining bright! As I looked at it, I felt that it was the first time, and only time, in my scientific career that I knew a law of nature that no one else knew. [...] It was the first time that I discovered a new law, rather than a more efficient method of calculating from someone else's theory (as I had done with the path integrals method for Schrödinger's equation and the diagram technique in quantum electrodynamics) [...] This discovery was completely new, although, of course, I learned later that others had thought of it the same time or a little before, but that did not make any difference.[9]

While the history of Feynman's works on weak interactions does not concern us here, the quotation is relevant in that it shows what Feynman thought (years later) about his diagrammatic methods.[10] In the end, the diagrams were not interpreted as a theory that could describe (either completely or partially) the reality of quantum phenomena, thereby drawing Feynman's view closer to Dyson's.

What remains of the physical reasoning and intuition behind Feynman's theorizing? At the end of the Nobel lecture, Feynman characterizes the physical reasoning as helpful to some people, but the only true physical description comes from the mathematics describing the experimental observations:

> The only true physical description is that describing the experimental meaning of the quantities in the equations —or bet-

---

[8]R.P. Feynman, Interviews and conversations with Jagdish Mehra, in Pasadena, California, January 1988.

[9]I believe that Feynman here was uncharitable to his own results (path integrals and Feynman diagrams) and I am in good company thinking it, for example: "[...] Richard, with his great talent for working out, sometimes in dramatically new ways, the consequences of known laws, was unnecessarily sensitive on the subject of discovering new ones. [...] Thus it would have pleased Richard to know (and perhaps he did know, without my being aware of it) that there are now some indications that his PhD dissertation may have involved a really basic advance in physical theory and not just formal development. " [27, 51]

[10]The quote above refers also to the path integrals method, which is strongly connected to Feynman diagrams.

ter, the way the equations are to be used in describing experimental observations. This being the case perhaps the best way to proceed is to try to guess equations, and disregard physical models or descriptions. For example [...] Dirac obtained his equation for the description of the electron by an almost purely mathematical proposition. A simple physical view by which all the contents of this equation can be seen is still lacking.

# 4 Conclusion

In this contribution I have explored the novel approach to quantum electrodynamics developed by Feynman with his famous diagrams. I argued that the novelty of the approach was exemplified by its rejection at the Pocono conference, where the graphical representation of scattering phenomena was ill-received by the audience because of lack of a rigorous mathematical formulation. I argued that Feynman's presentation was indicative of the physicist's beliefs that the diagrams can be considered as a partial representation of the depicted phenomena. I have then characterized the notion of partial representation through Boltzmann's Bildtheorie and as conveying explanatory phenomenological narratives.

In the second part of the contribution, I have maintained that Feynman's novel method can be traced back to some of his previous works, namely, the absorber theory of radiation and his interpretation of quantum mechanics. The absorber theory of radiation provided a new narrative with respect to classical electrodynamics, which is based on the direct-action-at-a-distance between particles, and on advanced and retarded radiation. I emphasized how this new narrative conveyed an alternative physical interpretation to the electrodynamics phenomena.

As a second example, I have addressed Feynman's dissatisfaction with the derivation of the Dirac equation from the attempt to quantize the absorber theory of radiation. The reason for the dissatisfaction was due to the lack of a clear physical interpretation of the mathematical results. This led to the formulation of the so-called quivering electron, a model that describes the relativistic electron as moving to the right or to the left on a lattice. Since it is unreasonable to think of Feynman as believing the denotative character of such a representation of the moving electron, I suggested to interpret the theory of the quivering electron in terms of partial representation —and thus as providing a narrative that conveys a physical interpretation which ought not be considered as a one-to-one representation of the physical phenomena.

Finally, I moved back to Feynman diagrams and further explained in what sense the visualization of the scattering processes fits Boltzmann's Bildtheorie. Along the lines of the absorber theory of radiation and of the quivering electron, Feynman diagrams should be conceived of as partial representations of scattering processes, without a demand for denotation, and by interpreting them a providing a phenomenological narrative.

In the last part of the contribution, I suggested that Feynman, later on in his career, drifted from his beliefs about the representational character of the diagrams and aligned more to the view maintained originally by Dyson. If Feynman, originally, believed that the diagrams were partial representations of physical phenomena, he then supported the view that only mathematics can provide a true description of phenomena, and that his diagrammatic method was but a useful tool for calculating scattering amplitudes.

# References

[1] Boltzmann, Ludwig (1974). Theoretical Physics and Philosophical Problems. Reidel.

[2] Feynman, R. P. (1949). Space-time approach to quantum electrodynamics. Physical Review, 76(6), 769.

[3] Wise, M. N. (2011). Science as (historical) narrative. Erkenntnis, 75, 349-376.

[4] Morgan, M. S. (2001). Models, stories and the economic world. Journal of Economic Methodology, 8(3), 361-384.

[5] Schweber, S. S. (1986). Feynman and the visualization of space-time processes. Reviews of Modern Physics, 58(2), 449.

[6] Dyson, F. J. (1949). The radiation theories of Tomonaga, Schwinger, and Feynman. Physical Review, 75(3), 486.

[7] Dyson, F. J. (1949). The S matrix in quantum electrodynamics. Physical Review, 75(11), 1736.

[8] Kaiser, D. (2009). Drawing theories apart. In Drawing Theories Apart. University of Chicago Press.

[9] Mehra, J., & Kragh, H. (1994). The beat of a different drum: The life and science of Richard Feynman. American Journal of Physics, 62(12), 1155-1155.

[10] Wheeler, J. A., & Feynman, R. P. (1945). Interaction with the absorber as the mechanism of radiation. Reviews of modern physics, 17(2-3), 157.

[11] Wheeler, J. A., & Feynman, R. P. (1949). Classical electrodynamics in terms of direct interparticle action. Reviews of modern physics, 21(3), 425.

[12] Feynman, R. P. (1966). The development of the space-time view of quantum electrodynamics. Science, 153(3737), 699-708.

[13] Forgione, M. (2020). The philosophical underpinning of the absorber theory of radiation. Studies in History and Philosophy of Science Part B: Studies in History and Philosophy of Modern Physics, 72, 91-106.

[14] Forgione, M. (2022). Feynman's space-time view in quantum electrodynamics. Studies in History and Philosophy of Science, 93, 136-148.

[15] Hartmann, S. (1999). Models and stories in hadron physics.

[16] Feynman, R. P. (1948). Space-time approach to non-relativistic quantum mechanics. Reviews of modern physics, 20(2), 367.

[17] Wüthrich, A. (2018). The exigencies of war and the stink of a theoretical problem: Understanding the genesis of Feynman's quantum electrodynamics as mechanistic modelling at different levels. Perspectives on Science, 26(4), 501-520.

[18] Schrodinger, E. (1930). Uber die Kraftefreie Bewegung in der relativistishen Quantenmechanik. Sitzung Phys.-Math., 31, 418.

[19] Breit, G. (1928). An interpretation of Dirac's theory of the electron. Proceedings of the National Academy of Sciences, 14(7), 553-559.

[20] Wüthrich, A. (2010). The genesis of Feynman diagrams (Vol. 26). Springer Science & Business Media.

[21] Stöltzner, M. (2018). Feynman diagrams: Modeling between physics and mathematics. Perspectives on Science, 26(4), 482-500.

[22] Forgione, M. (2020). Path Integrals and Holism. Foundations of Physics, 50, 799-827.

[23] De Regt, H. W. (2017). Understanding scientific understanding. Oxford University Press.

[24] De Regt, H. W. (1999). Ludwig Boltzmann's Bildtheorie and scientific understanding. Synthese, 119, 113-134.

[25] Stöltzner, M. (1999). Vienna Indeterminism: Mach, Boltzmann, Exner. Synthese, 119(1-2), 85-111.

[26] Brown, J. R. (2018). How do Feynman diagrams work?. Perspectives on Science, 26(4), 423-442.

[27] Gell-Mann, M. (1986). Dick Feynman—the guy in the office down the hall. Physics Today, 42(2), 50-54.

# Realism, underdetermination, and inference in cognitive neuroscience

Davide Coraci, Gustavo Cevolani
*IMT School for Advanced Studies Lucca*
davide.coraci@imtlucca.it, gustavo.cevolani@imtlucca.it

**Abstract.** In this paper, the realism/antirealism debate is discussed within the context of cognitive neuroscience, with a particular focus on experimental work that utilizes functional magnetic resonance imaging (fMRI). We start with a general discussion of realism and antirealism in cognitive science. Then, we consider the recent debate surrounding reverse inference, a reasoning pattern utilized by neuroscientists to support hypotheses about the engagement of cognitive processes based on patterns of brain activity observed through fMRI. We argue that reverse inference poses a problem of empirical underdetermination with respect to the theoretical interpretation of experimental results in cognitive neuroscience. As a result, it seemingly presents an antirealist argument regarding the evaluation of cognitive hypotheses. However, a closer examination of recent literature reveals that neuroscientists are successfully addressing the problem of reverse inference in a variety of ways that are closely related to how a philosophical defense of realism in the face of underdetermination would operate. These findings provide evidence for a qualified, realist interpretation of current cognitive neuroscience methodology.

**Keywords:** scientific realism; empirical underdetermination; reverse inference; fMRI; theoretical virtues; Bayesianism; abduction; cognitive neuroscience.

---

The authors wish to thank Sara Dellantonio and Luca Tambolo for helpful discussion, and two anonymous reviewers for their comments on the manuscript. Financial support from the Italian Ministry of Scientific Research (PRO3 project "Understanding public data: experts, decisions, epistemic values", CUP D67G22000130001) is gratefully acknowledged.

# 1 Introduction

Realists assert that it is justifiable to commit to our best scientific hypotheses, and to believe in both observable and unobservable aspects of the world as described by these hypotheses. Antirealists, on the other hand, disagree with this optimistic view and put forth several arguments to undermine it.

The philosophical literature exhibits a full range of such arguments and counterarguments, with the problem of empirical underdetermination playing a central role among others. This paper discusses the realism/antirealism debate within the context of cognitive neuroscience, with a particular focus on experimental work utilizing functional magnetic resonance imaging (fMRI). Section 2 reviews some well-known debates in cognitive science to illustrate how a broadly realistic view arguably underlies much current research in this field. Section 3 focuses on the recent debate concerning the issue of "reverse inference," interpreting it as an instance of empirical underdetermination, possibly favoring an antirealist view of neuroscientific methodology. Against this impression, Section 4 demonstrates how both neuroscientific and philosophical discussion of reverse inference appear to favor a realistic interpretation of scientific practice in this field. Two examples of such discussion are analyzed, highlighting how they closely parallel traditional defenses of scientific realism. A brief, tentative conclusion is offered in Section 5.

# 2 Realism and antirealism in cognitive science

Scientific realists argue that we are justified in committing towards our best scientific hypotheses in a rather strong sense, maintaining "a positive epistemic attitude towards the content of our best theories and models, recommending belief in both observable and unobservable aspects of the world described by the sciences" [9].

Such general attitude translates into more particular theses, depending on the specific variety of realism among the many currently discussed. Three of such theses, however, seem to be shared (in more or less qualified versions) by realists of all kinds [1, 29]. The first is the ontological thesis that the world that scientific theories aim to describe exists independently of our minds. The second is the epistemological thesis that knowledge of such mind-independent reality is indeed possible, being distilled in scientific theories and hypotheses that are at least approximately true representations of fragments of the world. The third is the methodological thesis that the methods adopted by scientists in their routine work do provide a means to produce and accumulate such knowledge. While contemporary scientific antirealists tend to share the ontological thesis, they usually object both to the epistemological and the methodological one.

Given the very general nature of the above mentioned three theses, they apply in principle to any scientific discipline, including cognitive science and neuroscience in particular.[1] Still, as far as we know, a discussion of the realism/antirealism debate in this area is lacking at the moment.[2] One of the goals of the present paper is advancing such discussion. We don't aim at a comprehensive reconstruction of the possible realist and antirealist arguments that have been and could be advanced relative to the theories and methodologies underlying cognitive neuroscience; this would require a much longer and detailed work. Instead, we take a look at the history of cognitive science in general to briefly discuss the ontological and epistemological realist theses as applied to cognitive science (in this section) and then focus on the methodological thesis with reference to cognitive neuroscience in particular (in the next section).

As we suggest, one can read many episodes in the historical development of psychology and cognitive science as instances of the realism/antirealism debate. One has to do with the very birth of the cognitive science research program. During the first decades of the last century, psychology was largely influenced by behaviorism (for discussion see, e.g., [3]). According to that research program, any reference to internal (and hence unobservable) mental states should be removed from the vocabulary of psychology in order to study cognition in purely observable (behavorial) terms. This radically empiricist view motivated a strong scepticism toward both the existence of mental entities and the possibility of scientifically investigate them. In this sense, behaviorism can be considered as an antirealist position, challenging both the ontological, the epistemological, and the methodological thesis of

---

[1] We follow here the traditional classification of neuroscience as one of the constitutive disciplines of cognitive science (along with philosophy, psychology, artificial intelligence, linguistics, and anthropology). For the purposes of this paper, we use "neuroscience" and "cognitive neuroscience" interchangeably, even if, strictly speaking, the latter is just a subarea of the former.

[2] As for neuroscience, this may well depend of the fact that this discipline is a young enterprise, dating back to the Nineties of the past century. Still, as we show in this section, the "pre-history" of cognitive neuroscience provides a number of interesting cues in the light of the realism/antirealism debate. For a quick presentation of the historical roots of contemporary neuroscience within traditional cognitive psychology and cognitive science, see [15]. For more general aspects see also [4, 23].

realism as applied to scientific psychology. As a consequence, one can interpret the "cognitive turn" that, between the Fifties and the Sixties, led to the birth of the classic cognitive science research program, as a realist response to the radically empiricist approach of behaviorism. Indeed, the crucial role played by mental representations and computations as the essential notions to investigate human and artificial minds apparently committed the newborn field of cognitive science to a more realist view of mental entities and cognition. Today, it seems fair to say that most cognitive scientists would subscribe to both the ontological and epistemological realist theses about mental entities and activities.

Of course, this is not to imply the the realism/antirealism debate in psychology and cognitive science has been settled in favor of the former position. Not only the controversy between cognitivists and behaviorists or neo-behaviorists continues today (see [28]), but also many other philosophical discussions over the past decades can be taken to represent a continuum of realist and antirealist positions about the nature of mental entities. For instance, the discussion about the existence of psychological states, especially those posited by folk psychology, provides a showcase of more or less realist positions, from dualism and epiphenomenalism (e.g., [26]), to functionalism and classical representationalism (e.g., [21]), reductionism, physicalism, and even more radical forms of eliminativism [11] (for discussion see also [42, 18]). A similar debate concerns the ontological status of phenomenal concepts or *qualia*: antirealists will maintain that *qualia* are mere illusions (e.g., [19]), or should be reduced to their representational and physical properties (e.g., [44]) or even eliminated (e.g., [12]), while realists tend to conceive *qualia* as irreducible and non-physical entities (see [10] for a general discussion).

Moreover, some developments within cognitive science have direct relevance to the realism/antirealism debate also in a different way. One example is the ongoing discussion about the "cognitive (im)penetrability" of perception [37]. A traditional view in both philosophy and psychology clearly distinguishes between perceptual experience, on the one hand, and cognition on the other hand. In this view, perception provides the real foundation for our knowledge of the external world: in particular, it is what we perceive that influences what we know about the world, and not the other way round. In this sense, human perceptual experience should be considered as independent from believes, desires, intentions, and cognition in general: in a slogan, perception is "cognitively impenetrable". This traditional view has been challenged by different authors (e.g., [13, 40]), who argue that our perceptual experience, such as, for instance, colour perception, is importantly affected both by cognitive states and by other theoretical constructs (e.g., colour constancy) on which our cognitive system systematically relies. The debate about cognitive penetrability of perception has relevant consequences for different epistemological questions. First, these results put under scrutiny the role of observation as foundation for knowledge [43] and question the objectivity of empirical knowledge as evidence used to test scientific theories. Indeed, if perception is cognitively penetrated, the concepts we use to describe the external world might not represent it in a way that is objective and reliable, leading to a repertoire of conceptual tools that appears too epistemologically weak to inform our scientific theories. Second, as discussed by several scholars (e.g., [7, 5]), if observations are cognitively penetrated, that seems to point towards some strong claim about the theory-ladenness of empirical data and the incommensurability of scientific theories, in favor of an antirealist perspective.

Those mentioned are only few examples of the ontological and epistemological questions that may arise in cognitive science. In retrospective, it seems fair to say that, as far as the cognitivistic approach has become the dominant view in the field, a realist view of the ontological commitment of neuroscientific and psychological theories towards the unobservable entities posited by their hypotheses is the dominant one, while strongly antirealist positions seem less viable (for instance [18, 5]). In this connection, the debate seems to have moved from the ontological and epistemological level to the methodological one. In other words, assuming that unobservable cognitive entities and processes exist and can be accessed by scientific investigation, how reliable are the methods and techniques employed by cognitive (neuro)scientists in producing robust knowledge about them? In the next section, we address this question discussing some recent debates concerning experimental work based on fMRI in the field of cognitive neuroscience.

# 3 Reverse inference and empirical underdetermination

Since the last decade of the 20th century, neuroimaging techniques started having a greater and greater impact on the field of cognitive science. A remarkable role was played by fMRI, a non-invasive technique for studying the brain by analysing changes in the cerebral oxygenated blood flow as a proxy for neural activity. A crucial goal of fMRI-based experiments is to establish associations between the engagement of cognitive processes $Cog$ (e.g., language, memory) and the activation of specific regions $Act$ in the brain, as observed in fMRI data. Among the different inferential strategies employed by neuroscientists for reasoning about *Cog-Act* associations, reverse inference received increasingly critical attention, starting with the analysis advanced by neuroscientist Russell Poldrack in 2006 [33].

To illustrate, suppose we are interested in confirming the hypothesis advanced by some previous study that a certain brain region *Act*, namely, the *pars opercularis* within the inferior frontal gyrus of the left hemisphere of the human brain[3], has a pivotal role in language processing (*Cog*). Suppose further that, to test that hypothesis, we run an fMRI-based experiment involving a linguistic task on a group of subjects and the acquired brain data show a significant activation of *Act* (i.e., an increased average activation relative to a control group). Now, given that evidence and the hypothesis from previous literature about the *Cog-Act* association, the researchers might be confident in inferring the recruitment of *Cog* (viz. language processing) from the activity of *Act* (viz. the pars opercularis). In other words, they seem justified in performing the following reasoning (cf. [33]):

(P1) In the fMRI literature, when the cognitive process *Cog* was putatively engaged under the experimental task $T$, the brain area *Act* was active.

(P2) In the present fMRI study, under the experimental task $T$, the activation of brain area *Act* is observed.

(C) Therefore, in the present study, the cognitive process *Cog* under task $T$ is engaged.

or, more formally:

(P1) $Cog \rightarrow Act$

(P2) $Act$

(C) $Cog$

Of course, this inference does not allow us to derive (C) with certainty, even when (P1) and (P2) are true. Indeed, from a logical viewpoint, reverse inference is a deductively invalid form of reasoning and, in particular, an instance of the "fallacy of affirming the consequent" [33]. This plain fact, by itself, is no decisive argument against the use of reverse inference (see [2] for critical discussion); still, it raises the question about the precise nature and

---

[3]Generally, when united to the region called *pars triangularis*, such area is known as Broca's area, from the pioneering studies of Paul Broca (1824-1880) investigating the role of this portion of the brain in elaborating language-related information.

role of such inferential pattern in neuroscientific studies. In this connection, the crucial issue is the lack of a unique, one-to-one association between $Cog$ and $Act$, i.e., between the activity of a certain brain region $Act$ and a unique cognitive function $Cog$. This well-known fact is known in the literature as the "problem of selectivity" of brain regions [33, 17]: in short, observing the activity of $Act$ is not a sufficient and necessary condition for concluding that the function $Cog$ was recruited. In fact, it is well-demonstrated from the fMRI literature that a variety of cognitive processes $Cog$ may be responsible of the activity of the same area $Act$, leading to a many-to-one relationship between them.[4] The status of reverse inference has raised much discussion in recent years, with neuroscientists and philosophers both criticizing its uncontrolled use to derive bold conclusions from fMRI data and defending the underlying reasoning pattern as ultimately sound if used with caution [16]. Without entering the debate, here we would like to introduce a new way of looking at the problem of reverse inference, which is instrumental to our main goal, i.e., discussing the issue of realism within cognitive neuroscience. From this perspective, we think it is useful to interpret the problem, and the discussion it raised, through the lenses of what philosophers of science are used to conceptualize as a problem of empirical underdetermination (for a related discussion see also [23]).

Empirical underdetermination amounts to the fact that, at least in principle, it is always possible to find two or more different (and possibly incompatible) scientific theories or models which account for a given body of data. In this sense, empirical evidence cannot "determine" its own theoretical interpretation: in the presence of "empirically equivalent" theories or models, the data are just insufficient to establish which of them is preferable on purely empirical grounds. This may happen both because two (or more) theories are empirically equivalent in the strong sense of being in principle compatible with any possible body of data in the relevant field, and hence essentially indistinguishable [1]. Or, more modestly, because they are, at least for the time being, compatible with the same available evidence, so that a choice between them seems prima facie unwarranted.

---

[4]By saying that one or more cognitive processes may be responsible for, and hence explain, the activity of a certain brain region, we are not endorsing a specific view of explanation (such as a causal conception), nor entering in the debate about reductionism in neuroscience. The debate on reverse inference is essentially methodological, and has developed in the literature quite independently from other discussions about the nature of neuroscientific explanation, the issue of causation, and that of the relations between the mind and the brain. We thank an anonymous reviewer for prompting us to make this point clear.

As an example of the second kind drawn from the literature on cognitive science, let us consider the well-known "imagery debate" [45, 31]. The debate concerns the specific format of mental representations that allow us to imagine real-world objects when we do not directly perceive them. According to the "imaginistic" or "analogical" perspective (e.g., [25]) mental images are represented by our cognitive system in a depictive or iconic format, i.e., they basically operate as real pictures in our mind. On the contrary, the "symbolic" account [38] takes mental images to have the same representational format of sentences, that is, a symbolic or propositional one, with no role left for a properly imaginistic representation. In short, the core of the discussion focuses on whether "the concept of image can be used as a primitive explanatory construct" [36] or should be further analysed by means of lower-level, symbolic representations. Interestingly, the empirical evidence available when the debate arose (for instance [39, 25]), mainly based on behavioral experiments, was compatible with both perspectives. In other words, it was hard to establish which of the two rival accounts, i.e., the analogical or the symbolic account, was preferable as a theory of the representation format of mental images.[5] In this sense, one can see the early imagery debate as a case of empirical underdetermination, where two competing theoretical accounts are empirically equivalent with respect to the available evidence, so that a rational choice between them is hard to justify on empirical grounds.

Going back to reverse inference, it seems to us that this kind of inference raises a clear case of underdetermination for hypotheses concerning the engagement of different cognitive processes. Of course, we are talking here of a quite modest form of underdetermination relative to low-level hypotheses of the form "in the present case, cognitive process $Cog$ is engaged". As said, the typical issue of reverse inference arises when two distinct cognitive processes $Cog_1$ and $Cog_2$ are both associated to a certain area $Act$ which is activated in a fMRI experiment. If our goal is explaining such activation by pointing to the engagement of one of the two processes, we face a problem of underdetermination. In other words, the schema introduced above can be revised as follows:

(P1) In the fMRI literature, when the cognitive process $Cog_1$ was putatively

---

[5] Nowadays, the debate appears resolved in favor of those advocating the depictive format of mental images. Indeed, as noticed by [31], there is strong evidence (based on neuroimaging works such as [32]) in favor of the thesis that humans represent information in multiple ways, with depictive images playing a key role in many areas of cognition.

engaged under the experimental task $T$, the brain area $Act$ was active.

(P2) In the fMRI literature, when the cognitive process $Cog_2$ was putatively engaged under the experimental task $T$, the brain area $Act$ was active.

(P3) In the present fMRI study, under the experimental task $T$, the activation of brain area $Act$ is observed.

(C) Therefore, in the present study, the cognitive process $Cog_1$ (instead of $Cog_2$) under task $T$ is engaged.

Since both $Cog_1$ and $Cog_2$ may be responsible of the activity of $Act$, $Act$ may equally support or confirm the recruitment of both $Cog_1$ and $Cog_2$, respectively, so making the conclusion in favor of (say) $Cog_1$ apparently untenable. Thus, as far as $Cog_1$ and $Cog_2$ are considered as competing neuroscientific hypotheses about the possible causes of $Act$, they are compatible with the same data and hence underdetermined by the available experimental evidence. In this sense, the problem of underdetermination – here depending on the problem of selectivity – seems intrinsic to the methodology of cognitive neuroscience as far as reverse inference is concerned. Whether such problem justifies a skeptical view of the neuroscientific quest for explanatory cognitive hypotheses of neural activity is discussed in the next section.

## 4 A realist look at the problem of reverse inference

In the last section, we argued that a problem of underdetermination lies at the methodological core of cognitive neuroscience, arising, in principle, each time reverse inference is employed to derive hypotheses about the engagement of cognitive processes from fMRI data. Since underdetermination figures prominently in many arguments against scientific realism in general, this motivates a reflection on whether the problem of reverse inference justifies an antirealist attitude about neuroscientific hypotheses and explanations. In this section, we argue that this is not the case.

Let us first recall how underdetermination seems to defy a realist view of scientific methodology in general. The intuition is that, if two competing theories or hypotheses are actually equivalent as far as the available evidence is concerned, a rational, evidenced-based choice between them must be unwarranted. In fact, given that this choice is underdetermined by empirical data, it is unclear what should justify a preference of one hypothesis over the other. In particular, if the two hypotheses differ in the unobservable constructs they posit (for instance, because they assume the engagement of different cognitive processes), one has no reason to think, on the basis of the available evidence, that one may be a better approximation to the truth than the other. In turn, this undermines the realist's claim that we should trust our better supported hypotheses as approximate representations of the target domain [9].

The general argumentative strategy just presented has long been used in the philosophical discussion to advance more specific arguments against different realist theses, often based on historical case studies. Of course, realists have reacted in a number of ways to each such critique, and the literature is now plenty of arguments and counterarguments surrounding the notion of underdetermination. Without trying to assess the debate, let us mention two of the main strategies employed by realists to resist antirealist conclusions from empirical underdetermination (for a more extended discussion, see [1]). First, one may doubt whether genuine cases of strong empirical equivalence of two actually different theories may exist, both from a conceptual and a historical point of view. If not, one may argue that theories that seem to be genuinely equivalent are actually two different versions of the same theory, so that "underdetermination does not look like a concrete problem in scientific practice, but a merely in-principle risk not to be taken too seriously" [1]. In that case, the antirealist argument based on underdetermination would be a non-starter. Second, even acknowledging that two genuinely different theories are compatible with the same evidence, this doesn't mean, the realist counter-objection goes, that they are empirically equivalent in any strong sense. In fact, the evidence in favor of a theory doesn't depend only on its empirical performance (so to speak), but also on the "theoretical virtues" enjoyed by theories, like for instance simplicity, high prior probability, explanatory power, and the like. If this is true, theoretical choice remains possible also in the face of empirical underdetermination. In sum, there is not shortage of realist answers to the challenge based on underdetermination, even if, as it is to be expected, antirealists are no less prolific in devising new objections.

Interestingly, the recent debate on the issue of reverse inference in cognitive neuroscience has apparently followed in the steps of the philosophical debate on underdetermination and its consequences for realism. This confirms our suggestions that looking at reverse inference as a problem of empirical underdetermination provides a useful entry point to the realism/antirealism debate within cognitive (neuro)science. To support this claim, we briefly discuss two cases in which the way reverse inference has been discussed is very close to how a defence of a realist stance in the face of underdetermination occurs in the philosophical literature.

**Reverse inference and Bayesian confirmation.** Starting with Poldrack [33], neuroscientists have resorted to a Bayesian analysis to address the problem of reverse inference and provide a conceptual analysis of such pattern of reasoning. Let consider, for instance, two competing cognitive hypotheses $Cog_1$ and $Cog_2$ and evidence $Act$ provided by the activation of some brain area as observed with fMRI. Proponents of the Bayesian analysis of reverse inference claim that the rational evaluation of $Cog_1$ and $Cog_2$ should be guided by their respective probability, as estimated via Bayes' rule. Thus, if the posterior probability of $Cog_1$ given $Act$, calculated as

$$p(Cog_1|Act) = \frac{p(Act|Cog_1)p(Cog_1)}{p(Act)} \qquad (1)$$

is greater than the posterior probability of $Cog_2$ (calculated in the same way as above), there is reason to prefer $Cog_1$ over $Cog_2$ as the most credible hypothesis accounting for the observed evidence $Act$. In this way, reverse inference is formalized as Bayesian reasoning, thus providing a possible way of choosing between $Cog_1$ and $Cog_2$ the best hypothesis given the available evidence $Act$.

Interestingly, the above proposal is in line with a classical argument in philosophy of science advanced by realists proposing a Bayesian response to the issue of empirical underdetermination. According to them, even if two competing hypotheses are empirically equivalent, it is still possible in general to differently assess them in Bayesian terms [22, 35, 1]. Indeed, even if the likelihoods of the two hypotheses — in our case, $p(Act|Cog_1)$ and $p(Act|Cog_2)$ — may be the same, expressing the fact that they are empirically equivalent, i.e., explain $Act$ equally well, their posteriors could well be different, if $Cog_1$ and $Cog_2$ differ in their prior probabilities $p(Cog_1)$ and $p(Cog_2)$. In general, such prior probabilities depend on how likely scientists deem the relevant hypotheses before any particular evidence is considered; in turn, this will depend on relevant background knowledge about the two hypotheses, including an assessment of their theoretical virtues as discussed in Section 4. In this way, a Bayesian analysis allows the realist to discriminate between empirically equivalent hypotheses, at least as far as their priors differ, thus overcoming the issue of empirical underdetermination.

In this connection, however, one should note that actual implementations of such Bayesian analysis of reverse inference in current neuroscientific practice cannot sustain such realist reading of reverse inference. The plain reason is that all proposals along these lines assume that the prior probabilities of the relevant hypotheses, for instance $p(Cog_1)$ and $p(Cog_2)$ are equal [33, 24].[6] This has motivated non-Bayesian proposals, like the one advanced by Edouard Machery [27], to treat reverse inference in purely "likelihoodist" terms, i.e., assessing hypotheses like $Cog_1$ and $Cog_2$ only in terms of their likelihoods $p(Act|Cog_1)$ and $p(Act|Cog_2)$. While defensible on their own, such proposals cannot of course be employed by realists to defend neuroscientific methodology in the face of underdetermination (nor this is their purpose); in fact, for empirically equivalent hypotheses it is sensible to assume that their likelihoods are the same, and hence cannot guide theoretical choice.

This, however, is not the end of the story. As we suggested elsewhere [14], one can defend a more sophisticated analysis of reverse inference, based on the notion of Bayesian confirmation, according to which the choice between $Cog_1$ and $Cog_2$ is not guided by a plain assessment of their posterior probabilities in the light of $Act$, but by the degree of empirical support or confirmation assigned to each hypothesis on the basis of $Act$. Such move allows one to differently evaluate empirically equivalent cognitive hypotheses by choosing the one with the highest degree of empirical support, even when one assumes that their prior probabilities are equal. This is especially true for those measures of Bayesian confirmation (like the so-called likelihood ratio) that only depend on the likelihoods $p(Act|Cog_1)$ and $p(Act|Cog_2)$ of the competing hypotheses, that can be empirically evaluated on a case-by-case basis. Interestingly, Poldrack himself proposes to use the so-called Bayes factors of competing hypotheses $Cog_1$ and $Cog_2$ in order to empirically compare them in the context of reverse inference. Since the Bayes factor, widely employed in a number of scientific disciplines, is a confirmation measure in the philosopher's sense, and in particular it is essentially equivalent to the likelihood ratio measure [41], this further illustrates how the current debate on reverse inference seems to move along the lines already traced by the realims/antirealism debate in the philosophy of science.

---

[6]Flat priors are assumed for different reasons. For instance, neuroscientists might have no specific expectation about the recruitment of a certain process during a task. Moreover, flat priors are generally assumed to avoid selection biases affecting the published literature. This assumption is especially true for the most widely employed tool for performing automated reverse inference based on large fMRI databases, i.e., NeuroSynth [46]. For a discussion of NeuroSynth in the light of such methodological issue see [15].

**Reverse inference as inference to the best explanation.** The Bayesian analysis outlined above is not the only way that philosophers and neuroscientists have tried to make sense of reverse inference and its use in experimental practice. Other scholars have proposed to interpret reverse inference as a form of abductive reasoning in the sense of Peirce (again, Poldrack himself mentions in passing such interpretation in his seminal paper [33]). According to such proposals [6, 34, 8], reverse inference can be viewed as an inference from the observed effect $Act$ to the engagement of some cognitive process $Cog$ as its putative cause. Following the terminology of [8], such pattern of reasoning can be employed both in a "weak" and in a "strong" form. In weak reverse inference, the conclusion about the engagement of $Cog$ is interpreted, in a more heuristic way, as a suggestion of a possible cognitive explanation of the observed activation, to be subjected to further exploration and testing. In its strong reading, instead, one views reverse inference as an "inference to the best explanation" (IBE) of the available evidence $Act$, which then confirms its conclusion $Cog$ as the most plausible candidate hypothesis accounting for the experimental results.

The discussion on the prospects and relative merits of weak and strong reverse inference is still open [8, 16]. To our purposes, it will be sufficient to note that abductive reasoning, and especially IBE, play a prominent role in the realism/antirealism debate, on at least two levels. First, the "no miracle argument", the central and perhaps more powerful argument in favor of scientific realism, can be construed as an abductive argument inferring, from the observed success of a scientific theory, to its approximate truth as the best explanation of such success [1, 9, 30]. Second, discussions of IBE crucially rely on the analysis of different theoretical virtues enjoyed by different hypotheses [20] and, as recalled in Section 4, such virtues play an important role also in realist arguments concerning underdetermination. Interestingly, IBE has recently entered the debate on reverse inference exactly along these latter lines.

As recent work has highlighted [8, 16], the evaluation of competing cognitive hypotheses in the light of available evidence from fMRI studies is not limited to the fact that the relevant hypotheses "fit the data", meaning they can account for the observed neural activations. Instead, other "virtues" of such hypotheses are assessed, like how well they cohere with currently accepted neuroscientific knowledge; how "good" they are as potential explanation of the evidence and how better they perform, under this respect, relative to their competitors; how able they are to unify different kinds of evidence (e.g., neural, behavioral, derived from animal models, etc.); and so on. Again, all such virtues have been carefully analyzed in the philosophical discussion of IBE, highlighting how simplicity, non-*adhocness*, coherence with background knowledge, greater explanatory power over closer competitors, unifying power, and the like all contribute in determining, first, what a "best explanation" is and, second, how reliable an IBE is in favor of the hypothesis best performing on such criteria [20]. In short, to the extent to which reverse inference can be construed as an instance of IBE and such kind of inference allows to overcome some problems connected with empirical underdetermination, the analysis just outlined suggests how a realist view of current scientific practice is tentatively warranted in the field of cognitive neuroscience.

# 5 Concluding remarks

In this paper, we looked at the issue of reverse inference in cognitive neuroscience through the lenses of the philosophical discussion of empirical underdetermination as a problem for a realist view of scientific inference. We first argued that the problem of reverse inference is essentially one of empirical underdetermination, thus challenging the methodological thesis of scientific realism, according to which the methods routinely employed by working scientists provide effective means to accumulate genuine knowledge about the relevant domain. Against this background, the critiques of reverse inference raised in the literature may apparently favor an antirealist view of neuroscientific methodology, even if leaving perhaps untouched the ontological and epistemological status of cognitive entities in general.

We then discussed two trends in the recent debate on reverse inference: one favoring a Bayesian analysis and the other considering reverse inference as IBE. In both cases, neuroscientists and philosophers have put forth arguments for favoring one cognitive hypothesis over others despite the problem of selectivity. As we suggested, these arguments amount to providing strategies for dealing with the issue of cognitive hypotheses which are empirically equivalent relative to the available neural evidence. Interestingly, the debate on reverse inference appears to closely track that on realism/anti-realism in general: the principal arguments in favor of reverse inference largely rely on theoretical virtues that competing but empirically equivalent hypotheses may possess differently.

In our view, this provides an intriguing instance of current scientific practice illustrating some more general philosophical principles at work. As we argued, the ongoing discussion on reverse inference, first, casts doubt on the notion that the problem of empirical underdetermination necessarily licenses an anti-realist attitude to (neuro)scientific methodology; second, it suggests that a realist view of reverse inference and related methodological problems is at least tenable, and in fact embedded in two central arguments favoring the employment of such inference.

# References

[1] Alai, M. (2017). The debates on scientific realism today: knowledge and objectivity in science. In: E. Agazzi (ed), *Varieties of Scientific Realism*, Springer, 19–47.

[2] Anderson, M. L. (2010). Review of *Neuroeconomics: Decision making and the brain*, *Journal of Economic Psychology* 31: 151–154.

[3] Bechtel, W., Abrahamsen, A., Graham, G. (2017). The life of cognitive science. In: G.Graham, W. Bechtel (eds),*A companion to cognitive science*, Blackwell Publishing Ltd., 1–104.

[4] Bechtel, W., Huang, L. T. (2022). *Philosophy of neuroscience*, Cambridge, Cambridge University Press.

[5] Beni, M. D. (2021). Cognitive Penetration and Cognitive Realism, *Episteme*: 1–16.

[6] Bourgeois-Gironde, S. (2010). Is neuroeconomics doomed by the reverse inference fallacy?,*Mind & Society* 9 (2): 229–249.

[7] Brewer, W. F., Lambert, B. L. (2001). The theory-ladenness of observation and the theory-ladenness of the rest of the scientific process, *Philosophy of Science* 68 (S3): S176–S186.

[8] Calzavarini, F., Cevolani, G. (2022). Abductive reasoning in cognitive neuroscience: weak and strong reverse inference, *Synthese* 200 (2): 1–26.

[9] Chakravartty, A. (2017). Scientific Realism. In: E. N. Zalta (ed), *The Stanford Encyclopedia of Philosophy*, Metaphysics Research Lab, Stanford University.

[10] Chalmers, D. J. (2006). Phenomenal concepts and the explanatory gap. In: T. Alter, S. Walter (eds), *Phenomenal Concepts and Phenomenal Knowledge: New Essays on Consciousness and Physicalism*.

[11] Churchland, P. M. (1981). Eliminative materialism and propositional attitudes, *The Journal of Philosophy* 78 (2): 67–90.

[12] Churchland, P. M. (1985). Reduction, qualia, and the direct introspection of brain states, *The Journal of Philosophy* 82 (1): 8–28.

[13] Churchland, P. M. (1988). Perceptual plasticity and theoretical neutrality: A reply to Jerry Fodor, *Philosophy of science* 55 (2): 167–187.

[14] Coraci, D., Cevolani, G. (2022). L'analisi bayesiana dell'inferenza inversa in neuroscienza: una critica, *Sistemi intelligenti* 34 (2):209–234.

[15] Coraci, D., Calzavarini, F., Cevolani, G. (2023). Reverse Inference, Abduction, and Probability in Cognitive Neuroscience. In: L. Magnani (ed), *Handbook of Abductive Cognition*, Springer.

[16] Coraci, D., Cevolani, G., Douven, I. *Inference to the Best Neuroscientific Explanation*, manuscript.

[17] Del Pinal, G., Nathan, M. J. (2017). Two kinds of reverse inference in cognitive neuroscience, *The human sciences after the decade of the brain*, Elsevier: 121–139.

[18] Demeter, T. (2009). Two kinds of mental realism, *Journal for general philosophy of science* 40 (1): 59–71.

[19] Dennett, D. C. (1988). Quining qualia, *Consciousness in contemporary science*: 42–77.

[20] Douven, I. (2022). *The art of abduction*, Cambridge (MA), MIT Press.

[21] Fodor, J. A. (1987). *Psychosemantics: The problem of meaning in the philosophy of mind*, MIT press.

[22] Glymour, C. (1980). Theory and evidence. In: Hanson, N. R., C W. Humphreys (eds), *Perception and discovery*, Springer.

[23] Hanson, S. J. E., Bunzl, M. E. (2010). *Foundational issues in human brain mapping*, Cambridge (MA), MIT Press.

[24] Hutzler, F. (2014). Reverse inference is not a fallacy per se: Cognitive processes can be inferred from functional imaging data, *NeuroImage* 84, 1061–1069.

[25] Kosslyn, S. M. (1980). *Image and mind*, Harvard: Harvard University Press.

[26] Lyons, J. C. (2006). In defense of epiphenomenalism, *Philosophical Psychology* 19 (6): 767–794.

[27] Machery, E. (2014). In Defense of Reverse Inference, *The British Journal for the Philosophy of Science* 65 (2): 251–267

[28] Nanay, B. (2019). Entity realism about mental representations, *Erkenntnis*: 1–17.

[29] Niiniluoto, I. (1999). *Critical scientific realism*, Oxford, Oxford University Press.

[30] Niiniluoto, I. (2018), *Truth-Seeking by Abduction*, Springer.

[31] Pearson, J., Kosslyn, S. M. (2015). The heterogeneity of mental representation: Ending the imagery debate, *Proceedings of the national academy of sciences* 112 (33): 10089–10092.

[32] Pearson, J., Naselaris, T., Holmes, E. A., Kosslyn, S. M. (2015). Mental imagery: functional mechanisms and clinical applications, *Trends in cognitive sciences* 19 (10): 590–602.

[33] Poldrack, R. A. (2006). Can cognitive processes be inferred from neuroimaging data?, *Trends in Cognitive Sciences* 10 (2): 59–63.

[34] Poldrack, R. A. (2011). Inferring Mental States from Neuroimaging Data: From Reverse Inference to Large-Scale Decoding, *Neuron* 72 (5): 692–697.

[35] Psillos, S. (1999). *Scientific Realism: How Science Tracks Truth*, London-New York, Routledge.

[36] Pylyshyn, Z. W. (1973). What the mind's eye tells the mind's brain: A critique of mental imagery, *Psychological bulletin* 80 (1): 1–24.

[37] Pylyshyn, Z. W. (1980). Computation and cognition: Issues in the foundations of cognitive science, *Behavioral and Brain sciences* 3 (1): 111–132.

[38] Pylyshyn, Z. W. (1981). The imagery debate: Analogue media versus tacit knowledge, *Psychological review* 88 (1): 16–45.

[39] Shepard, R. N., Metzler J. (1971). Mental rotation of three-dimensional objects, *Science* 171 (3972): 701–703.

[40] Siegel, S. (2019). Cognitive penetrability and perceptual justification, *Contemporary Epistemology: An Anthology*: 164–178.

[41] Sprenger, J., Hartmann, S. (2019). *Bayesian philosophy of science*, Oxford University Press.

[42] Sprevak, M. (2017). Realism about cognitive science. In: J. Saatsi (ed.) *Routledge Handbook of Scientific Realism*, London, Routledge: 357–368.

[43] Stokes, D. (2013). Cognitive penetrability of perception, *Philosophy Compass* 8 (7): 646–663.

[44] Tye, M. (1997). *Ten problems of consciousness: A representational theory of the phenomenal mind*, MIT press.

[45] Tye, M. (2000). *The imagery debate*, Cambridge (MA), MIT Press.

[46] Yarkoni, T., et al. (2011). Large-scale automated synthesis of human functional neuroimaging data, *Nature methods* 8 (8): 665–670.

# Can People Unlearn? A Reflection on the Conceptual and Cognitive Foundations of Organizations Systems Theory

Samuele Maccioni, Cristiano Ghiringhelli,
Edoardo Datteri
*University of Milano-Bicocca, "Riccardo Massa" Department of Human Sciences for Education*
`samuele.maccioni@unimib.it,`
`cristiano.ghiringhelli@unimib.it,`
`edoardo.datteri@unimib.it`

**Abstract.** Organizational science literature frequently employs biological metaphors, likening organizations to organisms that not only strive for survival but also learn from experience. Yet, this accumulated knowledge can become obsolete as internal and external environments evolve, necessitating the abandonment of outdated beliefs and knowledge—a process termed "unlearning." Introduced by Hedberg (1981) and Nystrom and Starbuck (1984), unlearning is defined as the deliberate discarding of old knowledge to make way for new insights, especially crucial in hypercompetitive environments. Despite its growing relevance, the concept of unlearning lacks a clear, consistent definition and its distinction from psychological concepts like forgetting remains unclear. Our study aims to clarify unlearning by achieving three objectives: delineating how unlearning is defined across organizational literature, exploring its relationship with psychological concepts, and proposing a cognitively plausible definition of unlearning. We propose a multidimensional taxonomy of unlearning, argue for its unique position within cognitive literature, and offer a definition that facilitates empirically testable theories. This work seeks to refine the theoretical foundation of organizational systems by elucidating a concept critical to adapting in rapidly changing environments.

**Keywords:** Unlearning, Organization's Theory, Taxonomy.

# 1 Introduction

Organizational sciences have increasingly looked towards biology for inspiration and guidance in their research and study [63]. One of the most prevalent metaphors used is to view organizations as organisms, much like living creatures that must learn from their experiences to survive [4]. Adopting this view due to changes in both the internal and external environment of an organization, previous knowledge and beliefs held by organizations may become outdated, incorrect, irrelevant or even misleading [39, 21, 59]. From an evolutionary perspective, the characteristics that in t0 allowed organizations to survive could be the same characteristics that in t1 led them to extinction.

In this connection, it is often claimed in the contemporary organizational literature that in order to not only survive but also thrive in hyper-competitive environments, organizations must continuously improve and *unlearn* so-called *path dependencies* [40]. Path dependencies are self-reinforcing mechanisms that anchor the organisation to the past. To illustrate, consider the constant use of the QWERTY keyboard even though the Dvorak keyboard, for example, has been universally recognised as more effective and efficient. Or how Kodak continued to ground its core business on film despite the advent of digital photography. This view is supported by recent studies that have identified unlearning as a crucial organizational ability [62, 13, 23, 71, 40]. These studies suggest that unlearning is a dual process that involves the acquisition of new knowledge as well as the conscious abandonment of outdated constructs [26, 47, 62]. Through this process, organizations can adapt to changing circumstances and remain competitive in their respective industries.

The concept of organizational unlearning was first introduced by Hedberg [26] and later by Nystrom and Starbuck [47]. By introducing the phenomenon to the organizational discourse, these articles laid out the foundational ideas that would have later informed all subsequent theorizing and reasoning on the role of unlearning in organizational science. Even though the phenomenon has been examined and scrutinized from diverse perspectives, these initial articles defined one of the fundamental pillars upon which the phenomenon is based: the interconnectedness between organizations and an environment that progressively becomes more hostile, demanding, and dynamic [14]. Furthermore, scholars posited that when organizations acquire new knowledge, a dual process occurs wherein knowledge becomes outdated at the same rate as it is learned due to changes in the external environment [7]. In this connection, it has been argued that a complete understanding of a subject or concept requires not only the acquisition of new knowledge but also the deliberate and conscious abandonment of outdated constructs. This process has been referred to in the literature as "unlearning". The phenomenon of unlearning has been generally construed as the intentional and mindful relinquishment of previously held beliefs, assumptions, and knowledge, leading one to reframe their understanding of, and perspective on, particular subjects or concepts. This notion of unlearning has since gained wide acceptance in organizational literature [53], particularly in recent years, as organizations navigate increasingly dynamic, unpredictable, and complex environments [46].

While the concept of unlearning has gained widespread use in the literature in recent years, it lacks a clear and precise definition, and existing definitions are often vague, imprecise, or contradictory [31]. The current body of literature on organizational unlearning has some gaps that require attention, and this work aims to address them in order to advance the theoretical foundations of this construct.

This article seeks to offer an overview of the various definitions and conceptualizations of unlearning found in organizational literature through a systematic review and analysis. It also aims to describe the connections between unlearning and related psychological concepts such as inhibition, interference, negative transfer, and forgetting [31], exploring their similarities, differences, and implications for organizational unlearning and change. Lastly, the paper endeavours to outline a cognitively plausible definition of unlearning, contributing to a more concrete and empirically testable understanding of the mechanisms involved.

By pursuing these three goals, this work intends to contribute to the understanding of a key concept in contemporary research on organizational systems and to possibly inform future research and practice in this area. We believe that this conceptual effort can have several payoffs even though the construct of unlearning, which characterized as unnecessary [31], is being continuously and increasingly used [46, 10, 34, 53]. This last fact does not in itself prove the robustness of the phenomenon, however, it denotes attention and interest from the academic community towards it. This. in our view, reinforces the need to strengthen the conceptualisation and empirical understanding of the concept.

# 2 Theoretical Background

The concept of unlearning, from the 1980s onwards [27, 26], has gained increasing attention in the academic debate [46]. Although the construct's origins can be traced back to the texts of Dewey [16], it is only in the 1980s-90s that the concept was introduced in the organisational literature, where it eventually flourished.

Unlearning, in the theoretical framework offered by Hedberg, Nystrom and Starbuck [27, 26, 47], refers to the intentional and conscious discarding of old organisational knowledge in a way that does not hinder the acquisition of new knowledge. According to their view, corroborated by experiments in the field of psychology [47], for new knowledge to be learnt effectively, it must replace obsolete knowledge. This idea is based on the assumption that, although organisations do not possess the counterpart of the "delete" key on computers [31, 65] organisations do possess a memory that can be "removed". However, the notion that both individuals and organisations can, and need to, "remove" previous knowledge or habits that inhibit the learning of novel knowledge and habits is not universally recognised and is the subject of a wide debate.

Other notions of unlearning have been proposed in the literature. Klein [36] defines the process of unlearning as the ability of organisations and individuals to replace old responses with new ones. [29] introduced a novel element into the debate, namely, the dimension of "challenge". According to this view, unlearning is a process in which existing cognitive structures, including dominant beliefs and values, particularly those of top managers, are challenged. In the debate on unlearning, while some authors have tried to give it a systemic definition capable of embracing the organisation and the parts of which it is made up, others, simplifying, have instead concentrated on exclusively organisational aspects, thus leaving out the cognitive part of the construct to focus instead on the processual one [4, 17, 62]. Their analyses were helpful not only to make the concept more manageable, but also to introduce the idea that unlearning is a process that precedes learning, and that aims at clearing the path from old knowledge or routines that inhibit the acquisition of new ones.

The phenomenon of unlearning has also been analysed along the cognitive dimension. [30] defines it as a conscious and deliberate process of reflection and preparation for the abandonment of existing knowledge, values and/or practices. This definition was later refined and extended by [19]. In particular, the two authors defined unlearning as a deliberate act of forgetting that implies a conscious decision to abandon knowledge, values and/or practices that organisations deem outdated and therefore no longer effective. Finally, other authors argue how unlearning occurs when the visions, attitudes and concepts one possesses are placed under the scrutiny of reflection to be recognised and subsequently rethought [43]. All these authors have apparently conflated unlearning with forgetting, which, however, does not involve the intentionality and awareness that characterise unlearning [34].

The discussion made so far highlights the fact that "unlearning" is a nuanced and multifaceted concept. Indeed, according to [32], unlearning should not be understood as a unitary and simple practice, but rather as the ability to gain and acquire alternative forms of knowledge and wisdom. Several authors, including [61], have pointed out that a comprehensive and unitary framework for understanding what organizational theory researchers talk about when they talk about unlearning is needed. Indeed, due to the lack of universally accepted conceptions of unlearning concerning both the construct and the process, the anecdotal evidence gathered from the community has made the process of understanding the characteristics concerning unlearning even more complex [9]. Therefore, although the concept of unlearning has been subjected to harsh criticism [31], the academic community seems to share a consensus regarding not only the need to unlearn but also the tracing of this process back to both the cognitive and behavioural spheres although it is impossible to gloss over the ongoing debate about how the process actually comes to life [53].

The scenario presented so far, composed of varied, intricate and different definitions which are difficult to disentangle, has exposed the concept to doubts and criticisms, and some scholars have even recommended that it be abandoned as a theoretical construct of organizational systems theory [31]. Even though previous works have made analogous remarks [36], we take this article [31] as an important step in the debate, as it provided compelling arguments for dropping any reference to unlearning. The main criticism made by the authors is that the concept of unlearning can be replaced by other, more familiar, psychological concepts without any loss of generality. Specifically, the phenomena that, according to the authors, can replace it, are *inhibition, interference, negative transfer* and *forgetting*.

One of the most debated overlaps in the literature is that between unlearning and forgetting [37]. However, unlearning cannot be equated with forgetting. The two differ from each other along a crucial dimension, namely, intentionality [34]. More specifically, forgetting is defined as the inability to recall something to mind that could have been remembered before instead, without this occurring intentionally. On the other hand, unlearning is defined as a deliberate process [62].

*Interference* typically refers to concurrent thoughts or processes that end up hindering one's performance, reducing its quality [50, 51, 69]. Interference, however, cannot be equated with unlearning. In this case, the two constructs cannot be overlapped as they identify different parts of the process called into question by unlearning. If anything, interference is a concept much closer to path dependencies [56]. To put it simply, interference is the impediment that must be overcome, and unlearning is the expedient that must be employed to do so.

The term "negative transfer" refers to the inhibition that stimulus-response processes impose on the acquisition of new information [22]. An effective example to understand how this phenomenon occurs can be found in the APA Dictionary of Psychology, which defines *negative transfer* as "a process in which previous learning obstructs or interferes with present learning. For instance, tennis players who learn racquetball must often unlearn their tendency to take huge, muscular swings with the shoulder and upper arm". This kind of interference inhibits learning in new contexts and cannot be equated to unlearning. Indeed, negative transfer denotes the object that unlearning is supposed to target as it could be seen also in the definition reported. Indeed, to be effective, unlearning processes must target factors that play a "negative transfer effect" by preventing organisations, groups, and individuals from growing or evolving [5, 33]. This said, the concept of negative transfer may be assigned a role in the definition of organisational unlearning, as it points to how old knowledge or habits can interfere with the acquisition of new information. *Negative transfer* also emphasizes the importance of identifying and discarding outdated knowledge, values, and practices that are no longer relevant or effective. Indeed, it is crucial to recognize when past experiences or habits are influencing current decision-making and hindering progress, therefore, by unlearning old introjections of stimulus-response processes, organizations can create space for new knowledge and behaviours that better align with their current goals and strategies.

Finally, the last phenomenon that is juxtaposed with unlearning is that of *inhibition*, which is defined in the literature as the intentional or unintentional blocking or overriding of a mental process [48, 58]. Unlike unlearning, *inhibition* phenomena involve a temporary decrease in the influence of certain information or processes on other processes. One key difference between inhibition and unlearning in an organizational context is their focus. Inhibition is primarily concerned with reducing or eliminating negative behaviours or practices while unlearning is concerned with exploring and adopting new, more effective approaches. Inhibition is often used as a short-term strategy to deal with immediate problems while unlearning is a long-term strategy for promoting ongoing growth and adaptation. Moreover, from a cognitive perspective, while inhibition refers to the suppression of pre-existing responses or behaviours, unlearning involves the modification or elimination of existing associations between stimuli or responses.

This reconstruction of the theoretical debate on the concept of unlearning highlights the multifaceted nature of the construct, and the fact that the same term is used in senses and with purposes that significantly differ from one another. However, this sort of theoretical confusion does not imply that the concept is useless and vacuous: quite on the contrary, it calls for an elaboration of a more stable and precise definition of it, also considering that, as argued so far and contra Howells and Scholderer [31], the phenomenon of unlearning cannot be easily equated with other more traditional psychological phenomena. Providing a satisfactory and unifying definition is out of the scope of this paper, which, however, aims at taking a first step towards this ambitious goal. The strategy adopted in the rest of the paper will involve systematizing the existing views on the subject, drawn from a systematic review of the literature from 1981 to the present day, in a three-dimensional taxonomy. Each dimension concerns a distinct aspect of the unlearning phenomenon: who does the unlearning? What is unlearnt? How does unlearning occur?

# 3 An Unlearning Taxonomy: Who?

Organizations are typically analysed at three levels of analysis: micro (individuals), meso (groups), and macro (organization) [67, 52, 15]. This approach is useful for several reasons. A multilevel perspective allows for a systemic approach to organizations, enabling one to observe organizational dynamics from different points of view and to identify interactions among them. At the micro level, one analyzes individual behaviours and their influence on the organization's well-being and productivity. On the meso level, the focus is on groups, on their structure, culture, communication, and decision-making processes. This allows for an understanding of how the organization consists of different political arenas [57] that interact with each other. Finally, at the macro level, one views organizations as organisms that move within a broader environment characterized by political, governmental, cultural, economic, and technological dynamics. This perspective allows for an understanding of external pressures that can modify the organization to adapt to its environment.

When it comes to defining the concept of unlearning, specifying who does the unlearning is clearly essential (under the assumption that unlearning is an activity, or a process, carried out by somebody or something). At least, it would be important to specify whether unlearning is a process carried out by individuals (micro-level), groups (meso-level), or entire organizations (macro-level). However, it's difficult to find clarifications of this sort in the literature most of the time, in fact, the dimension involved is not specified. In addition to this, it is interesting to note that two dimensions are often mentioned within the same definition. For example, Newstrom [45] without specifying the actor of the process defines unlearning as "the process of reducing or eliminating preexisting knowledge or habits that would otherwise represent formidable barriers to new learning". Yet, Becker [8] defines it as "the process by which individuals and organisations acknowledge and release prior learning" mentioning both the individual and organizational dimensions in the definition given.

This nuanced scenario is also returned by the distribution of definitions across the three levels. In fact, leaving out the vast majority of cases where the main actor of unlearning is not mentioned, we find that in the remaining cases, the dimension most frequently examined is the macro one, that is, which considers the *organization* as a whole as the main actor called upon by the process. Mehrizi and Lashkarbolouki [43] say that "unlearning refers to intentional practices organizations adapt to cope with their dependence on obsolete knowledge, processes and routines", or again Cegarra-Navarro and Wensley [12] refers to unlearning as the "organization's ability to prepare the ground for the creation and application of new knowledge".

In evidence of this, the *meso lens* turns out to be the least used. The group dimension turns out to be difficult to investigate at the organizational level since tracing the boundaries of a specific group is often problematic because of the cross-cutting and cross-functional processes that now characterize most organizations. Starbuck [55] refers to unlearning as "a process that shows people they should no longer rely on their current beliefs and methods", while Alas [2] says that during unlearning "people were expected to abandon their old ways of doing things".

Finally, it is interesting to note that the microlens embodying *individuals* is also little used within definitions, often even making it part of definitions that simultaneously refer to the organization as a whole as previously seen.

As we have seen, in most cases, the "who" dimension called into question by unlearning is not mentioned, or, when it is mentioned, it refers to the organisation as a whole. Identifying the "who" doing the unlearning with the entire organization is, in some sense, a convenient choice from a methodological point of view. Indeed, the memory of organizations, unlike that of individuals, can be traced physically, for example through documentation, procedures, or the know-how of figures placed in key positions. These elements just mentioned represent a large part of what is called organisational memory [66]. Since they are tangible elements, often even attested by documentation that is still on paper, they can be 'simply' removed through their physical elimination. Or, as in the case of top managers who perpetrate an obsolete way of doing things, they can be moved to another job or in extremis removed from the organisation. In light of this, it is therefore certainly easier to theorize and imagine an unlearning intervention than to do so by referring an individual to a group.

## 4 An Unlearning Taxonomy: What?

The second dimension identified here concerns the object of the unlearning process. For obvious reasons, one cannot understand what unlearning amounts to without understanding *what* is unlearnt. However, a wide variety of options can be found in the literature. More often than not, many objects are mentioned in the same definition.

Among the most frequently mentioned objects, we find *knowledge*. However, this term is often used without further qualification, making it difficult to gain a deeper understanding of what is really unlearnt. In particular, it is seldom specified whether the knowledge to be unlearnt is explicit or tacit. Other options include *routines, beliefs* and *values*. The mention of "routines" can be traced back to the landmark article by Tsang and Zhara [62], in which unlearning is defined as "The discarding of old routines to make way for new one, if any" (p.1437). But the list is not over yet: other definitions mention, *qua* objects of unlearning *processes, procedures, mental models, practices, methods*, and *norms*. The definition given by Matsuo [42] is clear proof of the dimensions called into question by unlearning, indeed in his view unlearning is "the changing of beliefs, norms, values, procedures, and routines to make way for new ones". Still, others mention *cognitive structures, habits* or *logic* [29, 1, 34].

What can be learnt from this analysis is that there is a substantial variety of views in the literature about what the object of unlearning is. The lack of convincing empirical evidence on the dynamics of unlearning processes may be regarded as a symptom of this variety. How can one study unlearning empirically, if there is no unitary view on what is unlearnt?

## 5 An Unlearning Taxonomy: How?

How does unlearning occur? How-questions can be addressed by identifying processes, or mechanisms carrying out processes. Not surprisingly, in light of the considerations made so far, we find a wide variety of views in the literature. The most frequently used term, used also in Tsang and Zhara's [62] article, is *discard*. "To discard" is defined in the Oxford dictionary as "to get rid of something that you no longer want or need." As this definition implies, to discard something is to throw it away, and this gives rise to theoretical issues. How can knowledge, routines, and all the other "whats" discussed in the previous section, be thrown away, completely removed, in a cognitive system? Moreover, one might reasonably claim that it is undesirable for organizations to completely throw away those "whats" - past experiences, although perhaps no longer fitting vis-à-vis the scenario, should be "reused" to make sense of new experiences and to analyse novel situations. Thus, equating unlearning with discarding leads one to a concept of unlearning that is not only cognitively implausible but also theoretically inappropriate.

Similar considerations can be made concerning other terms that are used in the unlearning literature, which notably include, among others, *eliminate*, *forget* or *clear out*. *Elimination* [45, 1] on the first hand alludes to the removal of outdated or irrelevant knowledge, practices or procedures from organisational memory. *To forget* [18, 4, 71], on the other hand, indicates an unintentional process in which organisations lose part of their knowledge (or some other kind of "what") over time. Clear out [42] finally involves a more systematic and thorough process of purging outdated or redundant knowledge, routines, and practices from the organisation.

All the terms discussed so far, despite their surface differences, allude to processes that (1) lead organizations to "throw away" the object of the unlearning process, and that, (2) in some cases, notably including "forgetting", are unintentional. Of a different nature are terms such as *change* [24, 44], *reflect* [30, 43], *question* [6] and *challenge* [29, 54, 68]. *Challenging* involves critically examining existing beliefs, assumptions and practices within the organisation. This process encourages individuals and teams to question the status quo and consider alternative ways of thinking and doing things. Moreover, *challenging* promotes a culture of continuous improvement, adaptability, and innovation by fostering an environment where it is safe to question and reevaluate existing norms. *Questioning* instead is the act of raising doubts, seeking clarification, or expressing curiosity about existing knowledge, routines, and practices. This cognitive mechanism may encourage open communication, critical thinking, and creative problem-solving. Additionally, when individuals and teams are empowered to ask questions, they can uncover and address hidden assumptions, biases, and inefficiencies that may be holding the organization back.

## 6  Towards a Definition of Unlearning

So far we have analysed and tried to rationally reconstruct the definitions of unlearning provided in the organizational system theory literature. We have emphasized the wide variety of views expressed by scholars in the field concerning the "who", the "what", and the "how" dimensions of unlearning processes. Even though providing a satisfactory and unitary definition of "unlearning" goes out of the scope of this article, the review made so far can orient the path towards this ambitious theoretical goal.

To try to compose it, we believe it is first necessary to focus on how the process occurs. Of the processes currently used to describe the process, we believe that the one most applicable and at the same time cognitively plausible is the one that adopts the challenging perspective [23, 54, 49, 68, 11]. Since old learnings cannot be intentionally forgotten or completely discarded [11] they must be proactively challenged. When we talk about "challenging", we refer to adopting a reflective posture that can systematically question the current way of doing things [6, 3, 43]. Moreover, the challenge dimension is functional for several reasons. First, it enables elements that often belong to an irrational and invisible sphere of organizations (such as routines and beliefs) to become visible and recognizable, and therefore open to being challenged [25]. Additionally, the dimension of the challenge has a clear beginning and end, allowing for timing, defining, monitoring, and evaluating the effectiveness of the unlearning process. Moreover, the challenge dimension directly involves the actors in the unlearning process deliberately and intentionally, drawing on their dynamic capabilities [20]. This fundamental and indispensable characteristic of unlearning [70] fosters participation and effectiveness of change [38].

Turning now to defining the elements that must be the object of the unlearning process we believe that the great common denominator uniting the objects of unlearning is their belonging to the past and the influence they now act on the present. In organizations, however, when we speak of these objects of unlearning, we are not only referring to the historical but also to all the data, processes and information [28] that feed the lenses through which the present is observed. The history and information that have ensured the survival of the organization risk in the present, however, anchoring them to the past through persistent, self-reinforcing mechanisms that are referred to in the literature as path dependencies [56]. These mechanisms, hindering the organization like barriers, generate negative transfer effects that by inhibiting the ability to learn prevent the organization from evolving and becoming [64]. As a result, if the ultimate objective of unlearning is to free an organization from aspects that keep it bound to an outmoded and misleading past [39, 59], a category into which we can bring all the taxonomic aspects identified in the what category, we argue that these elements may be gathered under the name of path dependencies. Thus seeing unlearning as challenging path dependencies such a definition can be applied to the entire organisational dimension.

We believe that this tentative proposal, albeit sketchy, has some advantages over the definitions discussed here. First, on the face of it, it identifies the phenomenon of unlearning in a relatively precise way, by taking a definite stance on the whos, whats, and hows of it. Even though its main terms (notably including "challenge" and "path-dependence") need further analysis, it circumscribes the phenomenon more neatly than the complex of positions discussed here. Second, it identifies a phenomenon of unlearning that is more cognitively plausible than phenomena that imply a complete discarding of knowledge and procedures. Third, it identifies a *peculiar* phenomenon, different in nature from, e.g., forgetting, the peculiarity being in its *intentional* nature. Whether this notion of unlearning can be helpful to the theoretical and empirical research on organizational systems, and be fully adequate from a descriptive and explanatory point of view, is a question to be addressed in future research. For the moment, we believe that the definition proposed here is, at least, one of the best candidates on the market.

# 7 Conclusions

As oxymoronic as it may seem to say, if in the modern scenario organizations were to look for a firm foothold to which they would still anchor themselves, they could not help but find it in change. Therefore, to stay abreast of the challenges imposed by the environment and contexts in which organizations are immersed, we can say that however confusingly it is still treated, the construct of unlearning is not only necessary but, once properly defined and structured, could prove to be a fundamental approach to be cultivated and applied. From the taxonomy carried out, as much as unlearning is a phenomenon so far theorized as a process in its own right [62, 60, 61, 10] it remains inescapably linked to its relative counterpart: learning. In its breadth, however, without overly forcing or circumscribing it, we cannot help but join other authors in necessarily considering it as an umbrella term capable of holding underneath the variety of constructs and phenomena addressed within the presented taxonomy. However, the current scenario should not relieve academics from attempting to give a definition anyway.

The definition we provide in the article, Organisational Unlearning as challenging negative transfer's path dependencies, attempts to reorder the constructs that have previously been called into question by scholars of the subject in order to stabilise a conception of unlearning that is aware of both the construct's potential and limitations. In reality, the taxonomy used allowed us to pick judiciously the aspects brought into question by our view of unlearning, intentionally specifying the players, objects, and processes in the issue. If one of the characteristics of unlearning is intentionality, we feel that the performed taxonomy, independent of the definition we present, may be a suitable example of a technique to use in order to bring order to the argument.

While we believe, however, that this definition, in addition to shaping itself as cognitively plausible, can try to bring order within the debate, we believe that there is still much work to be done around this concept in both theoretical and empirical terms. Therefore, we hope that this work can stand as a building block within a road that is still to be structured and travelled.

# References

[1] Akgün, A.E., Lynn, G.S., Reilly, R. (2002). Multi-dimensionality of learning in new product development teams. *European Journal of Innovation Management*, 5(2), 57–72.

[2] Alas, R. (2007). The triangular model for dealing with organizational change. *Journal of Change Management*, 7(3-4), 255-271.

[3] Antonacopoulou, E. P. (2009). Impact and scholarship: Unlearning and practising to co-create actionable knowledge. *Management Learning*, 40(4), 421-430.

[4] Argote, L. (2013). Organization learning: A theoretical framework. *Organizational learning: creating, Retaining and Transferring knowledge*, 31-56

[5] Atherton, M. (2007). A proposed theory of the neurological limitations of cognitive transfer. In *Annual meeting of the American Educational Research Association*, Chicago, IL (pp. 1-8).

[6] Baker, W. E., Sinkula, J. M. (1999). The synergistic effect of market orientation and learning orientation on organizational performance. *Journal of the academy of marketing science*, 27(4), 411-427.

[7] Bhatt, G. D. (2001). Knowledge management in organizations: examining the interaction between technologies, techniques, and people. *Journal of knowledge management*.

[8] Becker, K. (2005). Individual and organisational unlearning: directions for future research. *International Journal of Organisational Behaviour*, 9(7), 659-670.

[9] Becker, K. (2018). Organizational unlearning: time to expand our horizons?. *The Learning Organization*.

[10] Becker, K. (2019). Organizational unlearning: the challenges of a developing phenomenon. *The Learning Organization*.

[11] Brook, C., Pedler, M., Abbott, C., Burgoyne, J. (2016). On stopping doing those things that are not getting us to where we want to be: Unlearning, wicked problems and critical action learning. *Human Relations*, 69(2), 369-389.

[12] Cegarra-Navarro, J. G., Wensley, A. (2019). Promoting intentional unlearning through an unlearning cycle. *Journal of Organizational Change Management*.

[13] Cirnu, C. E. (2015). The shifting paradigm: Learning to unlearn. *Journal of Online Learning Research and Practice*, 4(1), 26926.

[14] Cousins, B. (2018). Design thinking: Organizational learning in VUCA environments. *Academy of Strategic Management Journal*, 17(2), 1-18.

[15] Czarniawska, B. (2017). Organization studies as symmetrical ethnology. *Journal of Organizational Ethnography*, 6(1), 2-10.

[16] Dewey J (1938) Experience and education. *Macmillan*, New York.

[17] De Holan, P. M., Phillips, N. (2004). Remembrance of things past? The dynamics of organizational forgetting. *Management Science*, 50(11), 1603-1613.

[18] Dodgson, M. (1993). Organizational learning: a review of some literature. *Organization Studies*, 14(3), 375-394.

[19] Durst, S., Zieba, M. (2019). Mapping knowledge risks: towards a better understanding of knowledge management. *Knowledge Management Research and Practice*, 17(1), 1-13.

[20] Eisenhardt, K. M., Martin, J. A. (2000). Dynamic capabilities: what are they?. *Strategic management journal*, 21(10-11), 1105-1121.

[21] Ethiraj, S. K., Levinthal, D. (2004). Modularity and innovation in complex systems. *Management Science*, 50(2), 159-173.

[22] Greeno, J. G., James, C. T., DaPolito, F. J. (1971). A cognitive interpretation of negative transfer and forgetting of paired associates. *Journal of Verbal Learning and Verbal Behavior*, 10(4), 331-345.

[23] Grisold, T., Kaiser, A. (2017). Leaving behind what we are not: Applying a systems thinking perspective to present unlearning as an enabler for finding the best version of the self. *Journal of Organisational Transformation and Social Change*, 14(1), 39-55.

[24] Gustavsson, B. (1999). Three cases and some ideas on individual and organizational re-and unlearning. In 6th *Workshop on Managerial and Organizational Cognition*:" Re-thinking in/by? Organisations/collectives". University of Essex, England.

[25] Haider, S. (2009). The organizational knowledge iceberg: an empirical investigation. *Knowledge and Process management*, 16(2), 74-84.

[26] Hedberg, B. (1981). How organizations learn and unlearn. *Handbook of organizational design* (1), 3-27.

[27] Hedberg, B. L., Nystrom, P. C., Starbuck, W. H. (1976). Camping on seesaws: Prescriptions for a self-designing organization. *Administrative Science Quarterly*, 41-65.

[28] Heikkila, E. J. (2011). An information perspective on path dependence. *Journal of Institutional Economics*, 7(1), 23-45.

[29] Hendry, C. (1996). Understanding and creating whole organizational change through learning theory. *Human Relations*, 49(5), 621-641.

[30] Hislop, D. (2013), Knowledge Management in Organizations: A Critical Introduction, 3rd ed., *Oxford University Press*, Oxford.

[31] Howells, J., Scholderer, J. (2016). Forget unlearning? How an empirically unwarranted concept from psychology was imported to flourish in management and organisation studies. *Management Learning*, 47(4), 443-463.

[32] Hsu, S. W. (2021). Exploring an alternative: Foucault-Chokr's unlearning approach to management education. *The International Journal of Management Education*, 19(2), 100496.

[33] Keung, Y. C., Ho, C. S. H. (2009). Transfer of reading-related cognitive skills in learning to read Chinese (L1) and English (L2) among Chinese elementary school children. *Contemporary Educational Psychology*, 34(2), 103-112.

[34] Klammer, A., Gueldenberg, S. (2019). Unlearning and forgetting in organizations: a systematic review of literature. *Journal of Knowledge management*.

[35] Klammer, A., Grisold, T., Gueldenberg, S. (2019). Introducing a 'stop-doing'culture: How to free your organization from rigidity. *Business Horizons*, 62(4), 451-458.

[36] Klein, J. I. (1989). Parenthetic learning in organizations: Toward the unlearning of the unlearning model. *Journal of management studies*, 26(3), 291-308.

[37] Kluge, A., Schüffler, A. S., Thim, C., Haase, J., Gronau, N. (2019). Investigating unlearning and forgetting in organizations: Research methods, designs and implications. *The Learning Organization*.

[38] Kraft, A., Sparr, J. L., Peus, C. (2018). Giving and making sense about change: The back and forth between leaders and employees. *Journal of Business and Psychology*, 33(1), 71-87.

[39] Leonard-Barton, D. (1992). Management of technology and moose on tables. *Organization Science*, 3(4), 556-558.

[40] Martignoni, D., Keil, T. (2021). It did not work? Unlearn and try again—Unlearning success and failure beliefs in changing environments. *Strategic Management Journal*, 42(6), 1057-1082.

[41] Matsuo, M. (2018). Goal orientation, critical reflection, and unlearning: An individual-level study. *Human Resource Development Quarterly*, 29(1), 49-66.

[42] Matsuo, M. (2019). Critical reflection, unlearning, and engagement. *Management Learning*, 50(4), 465-481.

[43] Mehrizi, M.H.R. and Lashkarbolouki, M. (2016), "Unlearning troubled business models: from realization to marginalization", *Long Range Planning*, Vol. 49 No. 3, pp. 298-323.

[44] Mezias, J., Grinyer, P., Guth, W. D. (2001). Changing collective cognition: a process model for strategic change. *Long range planning*, 34(1), 71-95.

[45] Newstrom, J. W. (1983). The Management of Unlearning: Exploding the" Clean Slate" Fallacy. *Training and Development Journal*, 37(8), 36-39.

[46] Nguyen, N. (2017). The journey of organizational unlearning: a conversation with William H. Starbuck. *The Learning Organization*.

[47] Nystrom, P., Starbuck, W. H. (1984). To avoid organizational crises, unlearn. *Organizational Dynamics*, p.53-66.

[48] Perfect, T. J., Moulin, J. A., Conway, M. A., Perry, E. (2002). Assessing the inhibitory account of retrieval-induced forgetting with implicit memory tests. *Journal of Experimental Psychology: Learning, Memory, and Cognition*, 28, 1111–1119.

[49] Rushmer, R., Davies, H. T. O. (2004). Unlearning in health care. *BMJ Quality Safety*, 13(suppl 2), ii10-ii15.

[50] Sarason, I. G., Stoops, R. (1978). Test anxiety and the passage of time. *Journal of consulting and clinical psychology*, 46(1), 102.

[51] Sarason, I. G., Sarason, B. R., Pierce, G. R. (1995). Cognitive Interference. In *International handbook of personality and intelligence* (pp. 285-296).

[52] Scott, W. R. (1995). Symbols and organizations: from Barnard to the institutionalists. *Organization theory: from Chester Barnard to the present and beyond*, 38-55.

[53] Sharma, S., Lenka, U. (2021). On the shoulders of giants: uncovering key themes of organizational unlearning research in mainstream management journals. *Review of Managerial Science*, 1-97.

[54] Sherwood, D. (2000). The unlearning organisation. *Business Strategy Review*, 11(3), 31-40.

[55] Starbuck, W. H. (1996). Unlearning ineffective or obsolete technologies. *International Journal of Technology Management*, 1996,ll: 725-737.

[56] Sydow, J., Schreyögg, G., Koch, J. (2009). Organizational path dependence: Opening the black box. *Academy of management review*, 34(4), 689-709.

[57] Thomas, R., Sargent, L. D., Hardy, C. (2011). Managing organizational change: Negotiating meaning and power-resistance relations. *Organization Science*, 22(1), 22-41.

[58] Toomela, A., Barros Filho, D., Bastos, A. C. S., Chaves, A. M., Ristum, M., Chaves, S., Salomão, S. J. (2020). Studies in the mentality of literates: 2. Conceptual structure, cognitive inhibition and verbal regulation of behavior. *Integrative Psychological and Behavioral Science*, 54(4), 880-902.

[59] Tripsas, M., Gavetti, G. (2000). Capabilities, cognition, and inertia: Evidence from digital imaging. *Strategic management journal*, 21(10-11), 1147-1161.

[60] Tsang, E. W. (2017). How the concept of organizational unlearning contributes to studies of learning organizations: a personal reflection. *The Learning Organization*.

[61] Tsang, E. W. (2017). Stop eulogizing, complicating or straitjacketing the concept of organizational unlearning, please. *The Learning Organization*.

[62] Tsang, E. W., Zahra, S. A. (2008). Organizational unlearning. *Human relations*, 61(10), 1435-1462.

[63] Tsoukas, H. (1991). The missing link: A transformational view of metaphors in organizational science. *Academy of management review*, 16(3), 566-585.

[64] Tsoukas, H., Chia, R. (2002). On organizational becoming: Rethinking organizational change. *Organization science*, 13(5), 567-582.

[65] Visser, M. (2017). Learning and unlearning: a conceptual note. *The Learning Organization*.

[66] Walsh, J. P., Ungson, G. R. (1991). Organizational memory. *Academy of management review*, 16(1), 57-91.

[67] Weick, K. E. (1976). Educational organizations as loosely coupled systems. *Administrative science quarterly*, 1-19.

[68] Wong, M.M.L. (2005) Organizational learning via expatriate managers: Collective myopia as blocking mechanism. *Organization Studies*, 26, 325-50.4

[69] Yee, P. L., Vaughan, J. (1996). Integrating cognitive, personality, and social approaches to cognitive interference and distractibility. *Cognitive interference: Theories, methods, and findings*, 77-97.

[70] Zahra, S. A., Abdelgawad, S. G., Tsang, E. W. (2011). Emerging multinationals venturing into developed economies: Implications for learning, unlearning, and entrepreneurial capability. *Journal of management inquiry*, 20(3), 323-330.

[71] Zhang, F., Lyu, C., Zhu, L. (2021). Organizational unlearning, knowledge generation strategies and radical innovation performance: evidence from a transitional economy. *European Journal of Marketing*.

# The problem of time for non-deparametrizable models and quantum gravity

Álvaro Mozota Frauca
*Autonomous University of Barcelona*
alvaro.mozota@uab.cat

**Abstract.** In this article I introduce a distinction between two types of reparametrization invariant models and I argue that while both suffer from a problem of time at the time of applying canonical quantization methods to quantize them, its severity depends greatly on the type of model. Deparametrizable models are models that have time as a configuration or phase space variable and this makes it the case that the problem of time can be solved. In the case of non-deparametrizable models, we cannot find time in the configuration or phase space of the model, and hence the techniques that allow solving the problem in the deparametrizable case do not apply. This seems to signal that the canonical quantization techniques fail to give a satisfactory quantization of non-deparametrizable models. As I argue that general relativity is non-deparametrizable, this implies that the canonical quantization of this theory may fail to provide a successful theory of quantum gravity.

**Keywords:** quantum gravity, problem of time, general relativity, canonical quantization.

## 1 Introduction

One of the strategies that physicists have deployed in order to build a theory of quantum gravity is to apply canonical quantization techniques to general

---

This research is part of the Proteus project that has received funding from the European Research Council (ERC) under the Horizon 2020 research and innovation programme (Grant agreement No. 758145) and of the project CHRONOS (PID2019-108762GB-I00) of the Spanish Ministry of Science and Innovation.

relativity. However, it is a well-known fact that the application of these techniques leads to what is known as the problem of time and, hence, that it affects all the approaches to quantum gravity that are based on this strategy[1]. In a nutshell, one follows the same quantization techniques that are used for quantizing gauge theories and what one obtains is that the supposedly physical states lack any temporal dependence. This has been known from the first quantization of general relativity by Wheeler and deWitt [4] and discussed since. Even though the two most comprehensive discussions and reviews of the problem in the 90s [12, 10] show some of the serious difficulties associated with this problem and the shortcomings of some of the resolutions proposed, the quantum gravity community has since adopted the view that the problem of time can be overcome in one way or another and that canonical quantization methods, when applied to general relativity, produce a meaningful theory, even if apparently timeless[2].

Some authors have been critical of this view[3], and in particular I have recently argued [14] that there are good reasons for believing that the canonical quantization of general relativity fails to give a satisfactory quantum theory. In this article I will expand on this criticism by leaving aside some of the more technical details and focusing on its more conceptual part.

I will start first in section 2 by introducing reparametrization invariance as a symmetry that physical models in general can have and that general relativity has in the form of its diffeomorphism invariance. This symmetry, or rather, the fact that it seems to be an essential feature of general relativity will be the source of the technical and conceptual problems associated with the problem of time. I will argue that this symmetry can be understood as a gauge symmetry, although there is an important difference with gauge theories like electromagnetism: while in these theories one can understand the transformation in a local fashion as leaving invariant the physical content at a time, this version of gauge transformations does not apply for reparametrization invariant theories.

In section 3 I will briefly introduce the canonical quantization procedure for gauge theories and I will explain the way in which it gives rise to the problem of time. That is, I will show that the quantum analogs of the constraint equations of any reparametrization invariant system automatically imply that the dynamical equation of the system, i.e., the Schrödinger-like equation, becomes trivial. This means that 'physical' states are time-

---

[1] Other approaches, such as string theory, are based on completely different ideas and techniques and avoid this problem.

[2] See for instance the discussions in [1, 11, 20, 21].

[3] See [3, 7, 8, 9].

independent, which constitutes the problem of time. I will also briefly discuss what I would count as a resolution to this problem.

Then, in section 4 I introduce the distinction between deparametrizable and non-deparametrizable reparametrization invariant models, which is crucial for my argument. A deparametrizable model is a model in which there is a variable in the configuration or phase space that can be identified with a time variable, or, alternatively, a model in which there is a set of variables in configuration or phase space that can be identified with spacetime coordinates. Conversely, a non-deparametrizable model is a model for which there isn't any such variable. I will give some examples of both kinds of models and I will argue that the problem of time can, in principle, be solved for deparametrizable models, while it is not the case for non-deparametrizable ones.

Finally, in section 5 I turn to the case of general relativity, I argue that it is a non-deparametrizable theory, and I therefore conclude that the problem of time cannot be solved for the canonical quantization of this theory. That is, contrary to what happens in the case of the deparametrizable models, the apparently timeless states one obtains by canonically quantizing the theory are indeed timeless and problematic.

## 2 General relativity and reparametrization invariant models as gauge theories

The problem of time arises because of the diffeomorphism of general relativity, and in general for any model with reparametrization invariance[4]. From a technical point of view, these models are expressed in the Hamiltonian form as constrained systems[5], similarly to gauge theories, and it is for this reason that one follows a similar quantization route. Moreover, from a conceptual point of view reparametrization invariance can be seen as a gauge symmetry, although in this section I will introduce some distinctions, contrary to some claims in the literature.

Let me start by defining reparametrization invariance. A model is reparametrization invariant if its solutions are expressed in terms of some param-

---

[4]In this sense, notice that the problem of time is more related to the fact that general relativity is generally covariant than with the fact that it is a gravitational theory. In the quantization of gravitational theories which are not generally covariant there wouldn't appear any problem of time.

[5]I refer the reader to [19] for an introduction to the Hamiltonian dynamics of constrained dynamics.

eters, typically some spacetime coordinates or some parameter parametrizing a trajectory or curve, and, given one solution, one can obtain a physically equivalent solution just by changing the parameters. More technically, the model defines solutions $f_A(\lambda_\alpha)$, where $f_A$ are the fields or variables in the theory, which vary with respect to a series of parameters $\lambda_\alpha$, and which are equivalent under invertible, differentiable transformations from one set of $\lambda_\alpha$ to another. These transformations are diffeomorphisms, although I will refer to them as reparametrizations and use the term diffeomorphism just when the parameters $\lambda_\alpha$ are coordinates on some spacetime manifold.

From the point of view of this article, the reparametrization invariance that will be relevant is temporal reparametrization invariance. That is, we will study systems that are described as Lagrangian systems in terms of evolution with respect to a 'time' variable or coordinate $\tau$ and which have a symmetry under invertible and differentiable transformations $\tau \to \tau'$. The case of general relativity is more complicated as there are also spatial diffeomorphisms and combinations of spatial and temporal diffeomorphisms, but for the purposes of this paper it is just temporal reparametrizations that will be playing a role.

Reparametrization invariance is similar in many aspects to the gauge invariance of theories like electromagnetism. Indeed, a reparametrization is a gauge symmetry in the sense that any two solutions of the equations of motion of the theory that are related by a reparametrization are physically equivalent, just as any two solutions of the equations of motion for the 4-potential of electromagnetism which are related by a gauge transformation are physically equivalent.

In a gauge theory like electromagnetism it is meaningful to speak about the observable quantities at a time $t$ because the gauge transformation doesn't affect the temporal structure of the theory and one can conceive of the gauge transformation as acting at a moment of time. For reparametrization invariant models this is not the case, as the action of the reparametrization is to change the physical state of the world associated with a given time parameter and one cannot interpret this as a transformation leaving unchanged the physical happenings at an instant defined by that time parameter. For instance, two diffeomorphism-related solutions of general relativity could assign the same time parameter $t = t_0$ to two radically different moments of the history of our universe, such as an instant just after the big bang and the instant in which I am typing these words. Clearly, the physical state of the world at such moments was completely different and it doesn't make sense to claim that it is the invariant content under a transformation relating both instants which captures the physical state of the world at time

$t_0$.

Arguments along these lines have been given in the literature[6] for reaching the same conclusion I am arriving at here, that reparametrization invariance is a gauge symmetry in the sense that it transforms solutions of the dynamical equations into equivalent solutions, but not in the more local sense, as the physical state at a coordinate time $t$ changes under such a transformation. This difference is relevant at the time of quantizing the theory, as it shows that imposing strict invariance under symmetry transformations would be physically nonsensical in the classical theory, but in the case of the quantum theory this conclusion seems not to be avoidable.

## 3  The problem of time

Given the conceptual and formal similarities between reparametrization invariant theories and standard gauge theories it is natural that one attempts to apply the same quantization techniques that work for the latter to the former. However, here I will show how this leads to a problem of time.

The canonical quantization method for gauge theories was first formulated by Dirac [5] and can be summarized as:

1. Start with a classical gauge theory defined on a phase space.

2. Choose a subalgebra of functions on phase space and quantize them, i.e., build an algebra of operators on a kinematical Hilbert space $\mathcal{H}_{kin}$ such that their commutator algebra is defined by the Poisson algebra of the classical functions.

3. Impose the quantum counterparts of the constraints $\phi_A$. That is, define the physical Hilbert space $\mathcal{H}_{phys}$ as the space of the states which satisfy $\hat{\phi}_A|\psi\rangle = 0$.

4. Build a Hamiltonian operator which is a quantization of one of the Hamiltonians that generate the constrained dynamics in the classical theory. The dynamics of the theory is contained in the Schrödinger equation for that Hamiltonian or in some equivalent form.

I refer the reader to [14] and the references therein for a detailed discussion of this quantization procedure and for the technical details. For the discussion here we just need to focus on steps 3 and 4 for the case of reparametrization

---

[6]See for instance [17, 18, 16, 14]. See also the more philosophical discussion in [13] in which this position is defended from the arguments in [6].

invariant models, as it is in the application of these steps that the problem arises.

In reparametrization invariant models, as in any gauge theory, there are some constraints, but unlike standard gauge theories, the Hamiltonian of the model is itself a constraint[7]. This fact is just a peculiarity in the classical case that doesn't prevent the definition of a consistent classical theory. In particular, it reflects the fact that the Hamiltonian plays not only the role of the generator of time evolutions but also the role of gauge generator in the global sense as described above. In the quantum theory, this peculiarity becomes problematic and gives rise to the problem of time.

Step 3 tells us that the constraints need to be imposed, i.e., that 'physical' states satisfy $\hat{\phi}_A|\psi\rangle = 0$. However, as the Hamiltonian is of the form $H = v^A \phi_A$ this leads to a trivial Schrödinger equation:

$$\partial_t|\psi\rangle = \hat{H}|\psi\rangle = 0. \qquad (1)$$

This means that 'physical' states are time-independent, contrary to our expectation from standard quantum theory in which states carry a temporal dependence that allows describing their evolution. In this sense, it seems that we have found a timeless theory in which nothing evolves with time. This is the problem of time for reparametrization invariant theories[8].

This is prima facie problematic because it seems that we are missing an important part of what we take to be a quantum theory, its dynamical aspect, and therefore it seems that the quantization has failed. However, in section 4 I will show how for a class of models, the deparametrizable ones, the problem can be solved in the sense that one is able to interpret states in the physical Hilbert space as dynamical states, i.e., as describing a temporal evolution and recovering the standard quantum-mechanical picture. For the rest of the models, the non-deparametrizable ones, I will argue that it is not the case and that the quantization indeed fails.

---

[7]I refer the reader to [14] for a detailed discussion of the constraints that appear in reparametrization invariant systems and of appropriate Hamiltonian for describing such systems. There are three possible Hamiltonians one can use for describing the system (the canonical, the total, or the extended) but I won't argue here for choosing one of them over the others as the relevant fact is that the three of them are constraints.

[8]For more extensive and detailed accounts see the classical reviews [12, 10] and the more recent book [1]. There are several more or less connected issues that are referred to as 'the problem of time', as is well noted in these references. Here I am focusing on what is also known as the frozen formalism problem, i.e., in the fact that the formalism obtained lacks temporal dependence.

Let me clarify that by 'solving' the problem of time I don't mean that one recovers a standard quantum theory, as one may very well suspect that the standard structures of quantum theory need to be modified when one attempts to formulate a theory of quantum gravity. A resolution of the problem of time, I argue, consists in showing that a meaningful theory can be built from the 'timeless' states obtained in the canonical quantization, which would allow us to claim that the quantization has not failed. This definition of resolution is quite open and it does not presuppose that a meaningful theory is just a theory in the form of a standard quantum theory. Even in this wide sense of resolution, I will argue next that for non-deparametrizable models we lack a successful resolution, and hence that the canonical quantization of these models fails.

## 4 Deparametrizable and non-deparametrizable systems

As I just mentioned, the main feature of a reparametrization invariant theory that determines whether the problem of time can be solved or not is whether it is deparametrizable or not. In this section I will give a definition and examples of both deparametrizable and non-deparametrizable systems and I will argue why the problem of time can be solved for the former but not for the latter.

Let me start by giving the definitions of both types of reparametrization invariant systems:

**Deparametrizable model:** A model is deparametrizable if one can identify a time variable or a set of spacetime coordinates among its configuration or phase space variables.

**Non-deparametrizable model:** A model is non-deparametrizable if it is not deparametrizable.

A deparametrizable model is one that can be deparametrized, i.e., it is possible to express evolution not with respect to the arbitrary parameter(s) of the reparametrization invariant model, but with respect to the time variable, or spacetime coordinates, that can be found among the variables of the model. For instance, consider the model given by the following action:

$$S[x,t] = \int d\tau \left(\frac{1}{2}m\frac{\dot{x}^2}{\dot{t}} - \dot{t}V(x)\right). \tag{2}$$

Solutions of the equations of motion for this action are of the form $x(\tau), t(\tau)$ and describe the trajectory of a body in Newtonian spacetime under the effect of the potential $V$. The trajectory is independent of the way it is parametrized and it can be deparametrized and expressed with respect to the true time variable $t$. That is, one can express the trajectory just as $x(t)$.

A necessary condition for deparametrization is that the variable to be identified as a time variable needs to be monotonic with respect to the arbitrary parameter $\tau$. However, it is not a sufficient condition, as even in models in which there are physical variables that are monotonic they can be argued to be different from a time variable. For instance, the position of a particle is monotonic in Newtonian physics if it is a free particle, but it is different from time as we can conceive a situation in which it is affected by some potential and ceases to be monotonic. In this sense, my definition of deparametrizable is not purely formal regarding monotonic functions, but has a component of how we interpret some variables as time variables or spacetime coordinates.

For the case of non-deparametrizable models there is no variable that plays the role of time (or spacetime coordinates) in the configuration or phase space of theory and one cannot perform a deparametrization. The most popular example of such a model is the Jacobi action for any Newtonian system[9]. As an example, let me introduce the Jacobi action for a system of two harmonic oscillators:

$$S[x,y] = 2\int d\tau \sqrt{\frac{m}{2}(\dot{x}^2+\dot{y}^2)\left(E-\frac{1}{2}(k_x x^2+k_y y^2)\right)}. \qquad (3)$$

Solutions of the equations of motion of this system are pairs $x(\tau), y(\tau)$ which describe the trajectories of the two oscillators with respect to an arbitrary temporal parametrization $\tau$. In figure 1 I represent two equivalent such parametrizations that represent the same physics, i.e., the same sequence of configurations. Note that neither of the configuration space variables is a time variable, as both have a different physical interpretation and furthermore none of them is monotonic.

To recover a preferred notion of time one can 'gauge transform' to Newtonian time $t$ by means of the relation $dt = \sqrt{\frac{m(\dot{x}^2+\dot{y}^2)}{2E-k_x x^2-k_y y^2}}d\tau$ that relates Newtonian time with the parameter $\tau$ describing the same trajectory using a different parametrization. In this sense, it is important to emphasize that the physical content of the model is encoded in the trajectories $x(\tau), y(\tau)$ or

---

[9]The relevance of this kind of model for the discussion of the problem of time was first noticed by Julian Barbour [3]. The discussion in [7] brought the topic again to the fore.

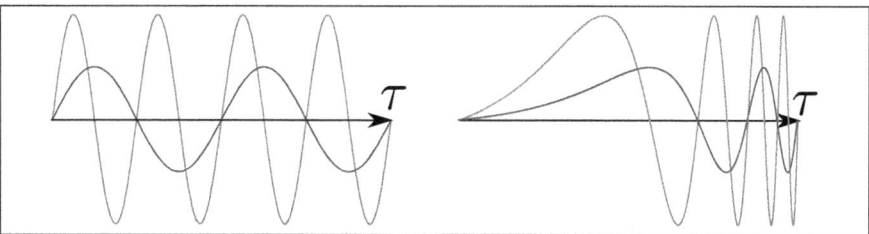

Figure 1: Two equivalent solutions of the equations of motion of the double harmonic oscillator model. They represent the same sequence of oscillations but they differ from each other at the particular values of $\tau$ they assign to each moment of time. In this way the parametrization on the left-hand side represents the Newtonian parametrization in which the oscillations are regular and the one on the right-hand side represents a parametrization in which the oscillations become faster as $\tau$ passes.

$x(t), y(t)$, but not on some 'deparametrized' $x(y)$ or $y(x)$. To insist, while in the deparametrizable case we could identify one configuration variable as time, this is not the case in the non-deparametrizable case.

This fact is determinant for the different ways the problem of time affects both kinds of models. For the case of deparametrizable models, states in the physical Hilbert space can be interpreted, in one way or another, as describing dynamical states with respect to a time variable and therefore the problem is solved. I refer the reader to [14] for a detailed discussion of the different strategies proposed for solving the problem in this case and for an argument showing that they all rely on a deparametrization. Here I will give just a simple account of the example above that can be generalized to more complicated systems and to different strategies of resolution.

Applying the quantization steps in section 3 to the deparametrizable system described by action 2 we are lead to a problem of time, as we find states of the form $\psi(x, t)$ which do not evolve with respect to $\tau$. However, this is not problematic as we can nevertheless interpret $\psi(x, t)$ not as a state at 'time' $\tau$ but as a wavefunction for the true configuration variable $x$ which evolves with respect to time $t$. Moreover, the quantum constraint equation for this system is just the Schrödinger equation and one can interpret states in the physical Hilbert space as solutions to the dynamical equation of the quantum system. In this sense, the problem of time can be solved for this system: the apparently timeless states were hiding just standard quantum mechanics.

However, the case for non-deparametrizable models is not so positive. Consider the case of the double oscillator above. 'Physical' states are states of the form $\psi(x,y)$ and, contrary to what happened in the deparametrizable case, they are clearly not the dynamical states one would expect from the non-reparametrization invariant form of the model, which would be states of the form $\psi(x,y,t)$ where $t$ is the Newtonian time variable. One could try to look for an interpretation that would nevertheless interpret $\psi(x,y)$ as states containing dynamical information and from which a meaningful theory could be devised. For instance, one could try to mimic the interpretation for the deparametrizable system and interpret $\psi(x,y)$ as a wavefunction on the configuration space for $x$ that evolves with respect to the 'time' $y$. However, this is problematic for several reasons. First, it breaks the symmetry between $x$ and $y$ as both of them play the same role in the classical theory and receive the same interpretation. Second, it seems wrong to identify $y$ as a time variable as it clearly wasn't a time variable in the classical theory and as it wasn't even monotonic. In the classical theory $y$ was oscillating in time and a value $y_0$ couldn't be used to identify a particular instant of the trajectory, as there were many, infinitely many indeed, instants in which the oscillator took the position $y_0$. In the quantum theory there would be just one wavefunction $\psi(x)$ for 'time' $y_0$ which wouldn't be reflecting the oscillating behavior of the oscillator. Furthermore, if we read $y$ as a time variable we would take $y \to -\infty$ as the distant past and $y \to \infty$ as the distant future in the quantum theory, while in the classical theory the values of $y$ were bounded and oscillating. Moreover, the relationship between the quantum theory we have obtained and the classical theory is unclear to say the least, as it seems that we cannot recover the behavior of the classical theory starting from the quantum one.

Similar objections can be raised to the different strategies for finding a resolution that have been proposed in the literature. That is, when applied to the model of the two harmonic oscillators, the strategies that work for deparametrizable models can be argued to fail. This is mainly because these strategies need in a more or less explicit way a deparametrization, and when it is not available one runs into trouble. I refer the reader to [14] for a detailed discussion of the different strategies and the way they seem to fail in the case of the double harmonic oscillator.

Let me also mention that one could try to devise a Bohmian strategy in which even if the state is time-independent this wouldn't be problematic if the dynamics of the Bohmian particles or fundamental entities is non-trivial. However, the guidance equation would need to be changed and I am unaware of any generic proposal for solving this for any reparametrization invariant system. There are some proposals for the case of the quantization of general relativity, but they aren't popular in the quantum gravity literature[10].

For these reasons, I have argued that the canonical quantization of non-deparametrizable models fails, or, at least, that the proposed resolutions for the problem of time for these models have serious shortcomings that would need to be addressed in order to consider that they show how the timeless states obtained by canonically quantizing these models could encode a meaningful theory.

## 5  Does the quantization fail for general relativity?

Finally, we can turn to the case of general relativity. As I have advanced before, I will argue that general relativity is not deparametrizable and therefore that the problem of time for its canonical quantization cannot be solved as it was solved in the case of deparametrizable models.

For the canonical quantization procedure, general relativity is expressed as a Lagrangian theory in which the geometry of space and matter evolve with respect to an arbitrary time parameter, that is, a foliation is introduced that splits the relativistic spacetime into space and time[11]. The degrees of freedom of the theory are a three-dimensional metric $g_{ab}$ that describes the geometry of space, the lapse function $N$ and shift vector $N^a$ that encode the time-like components of the metric, and possibly some degrees of freedom corresponding to matter. Under a diffeomorphism, these degrees of freedom are affected, but we know that the diffeomorphism-related description is just an equivalent description of the same spacetime. For instance, one can have two diffeomorphism-related descriptions of Minkowski spacetime, one in which time slices describe flat spaces and another one in which they don't.

A priori, all the variables in this description have a physical interpretation that is different from being a time variable or a spacetime coordinate. In this sense, if general relativity were deparametrizable, it would not be in a straightforward way. Some authors proposed some ways in which spacetime geometry could be encoding time in some way which could be used for claiming that general relativity is deparametrizable after all[12]. However, the

---

[10]See [15] and references therein.

[11]Indeed, not every relativistic spacetime can be decomposed in this way, and they are left out for the quantization of the theory.

[12]See for instance [2, 11].

consensus about this issue is that general relativity isn't deparametrizable, as it was argued for instance in [12]. One of the reasons for this is that it seems quite clear that there is no phase space function that satisfies the necessary condition of being monotonic. In [14] I further counter-argued against some arguments in the literature used for supporting the claim that general relativity is deparametrizable. My analysis of these arguments shows that they could be applied to the case of the double harmonic oscillator and that they would lead to the wrong conclusion that this system is deparametrizable. As these arguments aren't valid when applied to the case of the double harmonic oscillator model, I argued that they aren't valid when applied to general relativity.

There is a further argument linking the structure of general relativity to the structure of our example above. The definition of proper time in general relativity is analogous to the definition of Newtonian time in the model defined in the Jacobi action. In particular, proper time is defined by:

$$ds^2 = -g_{\mu\nu}dx^\mu dx^\nu = N^2 dt^2 - g_{ab}(dx^a + N^a dt)(dx^b + N^b dt), \quad (4)$$

which is just analogous to $dt = \sqrt{\frac{m(\dot{x}^2+\dot{y}^2)}{2E-k_x x^2-k_y y^2}} d\tau$ for the double oscillator. For these reasons, there are good grounds for comparing general relativity with the example of the double harmonic oscillator and arguing that if the quantization fails for one of the two, then the quantization must fail for the other.

This conclusion puts some pressure on the quantum gravity community, at least on those working on approaches based on canonical quantizations of general relativity. In particular, this affects approaches like quantum geometrodynamics, loop quantum gravity, loop quantum cosmology, and some other approaches to quantum cosmology. There are a couple of ways the community could react to defend these approaches: they could either deny that general relativity is a non-deparametrizable model, or argue that it is of a different kind from the example in this paper, so that the comparison doesn't affect their theories; or they could argue that there is a resolution of the problem of time that is satisfactory for both general relativity and other non-deparametrizable models. In any case, the defender of these approaches would need to clarify the way in which they are believed to constitute satisfactory quantum theories.

# References

[1] Anderson, E. The Problem of Time. *Fundamental Theories Of Physics*. **190** (2017)

[2] Baierlein, R., Sharp, D. & Wheeler, J. Three-dimensional geometry as carrier of information about time. *Physical Review*. **126**, 1864-1865 (1962,6)

[3] Barbour, J. The timelessness of quantum gravity: I. The evidence from the classical theory. *Classical And Quantum Gravity*. **11**, 2853 (1994,12)

[4] DeWitt, B. Quantum theory of gravity. II. The manifestly covariant theory. *Physical Review*. **162**, 1195-1239 (1967,10)

[5] Dirac, P. Lectures on Quantum Mechanics. (Belfer Graduate School of Science Yeshiva University,1964)

[6] Earman, J. Thoroughly Modern Mctaggart: Or, What Mctaggart Would Have Said If He Had Read the General Theory of Relativity. *Philosophers' Imprint*. **2** pp. 1-28 (2002)

[7] Gryb, S. Jacobi's principle and the disappearance of time. *Physical Review D*. **81**, 044035 (2010,2)

[8] Gryb, S. & Thébault, K. Symmetry and Evolution in Quantum Gravity. *Foundations Of Physics 2014 44:3*. **44**, 305-348 (2014,3)

[9] Gryb, S. & Thébault, K. Time remains. *British Journal For The Philosophy Of Science*. **67**, 663-705 (2016)

[10] Isham, C. Canonical Quantum Gravity and the Problem of Time. *Integrable Systems, Quantum Groups, And Quantum Field Theories*. pp. 157-287 (1993)

[11] Kiefer, C. Quantum GravityThird Edition. (Oxford University Press,2012,4)

[12] Kuchař, K. Time and interpretations of quantum gravity. *Proceedings Of The 4th Canadian Conference On General Relativity And Relativistic Astrophysics*. (1992,7)

[13] Maudlin, T. Thoroughly Muddled Mctaggart: Or, How to Abuse Gauge Freedom to Create Metaphysical Monstrosities. *Philosophers' Imprint*. **2**, 1-23 (2002)

[14] Mozota Frauca, Á. Reassessing the problem of time of quantum gravity. Gen Relativ Gravit 55, 21 (2023)

[15] Pinto-Neto, N. & Struyve, W. Bohmian quantum gravity and cosmology. *Applied Bohmian Mechanics.* pp. 607-664 (2018,1)

[16] Pitts, J. Equivalent theories redefine Hamiltonian observables to exhibit change in general relativity. *Classical And Quantum Gravity.* **34**, 1-23 (2017)

[17] Pons, J. & Salisbury, D. Issue of time in generally covariant theories and the Komar-Bergmann approach to observables in general relativity. *Physical Review D.* **71**, 124012 (2005,6)

[18] Pons, J., Salisbury, D. & Sundermeyer, K. Observables in classical canonical gravity: Folklore demystified. *Journal Of Physics: Conference Series.* **222**, 012018 (2010,4)

[19] Rothe, H. & Rothe, K. Classical and Quantum Dynamics of Constrained Hamiltonian Systems. (WORLD SCIENTIFIC,2010,4)

[20] Rovelli, C. Quantum Gravity. (Cambridge University Press,2004,11)

[21] Rovelli, C. & Vidotto, F. Covariant loop quantum gravity: An elementary introduction to quantum gravity and spinfoam theory. *Covariant Loop Quantum Gravity: An Elementary Introduction To Quantum Gravity And Spinfoam Theory.* pp. 1-254 (2015,1)

# Artificial Intelligence as Expected Intelligence

Martina Bacaro, Francesco Bianchini
*University of Bologna*
`martina.bacaro2@unibo.it, francesco.biachini@unibo.it`

**Abstract.** The inquiry into the nature of intelligence within artificial intelligence (AI) has been a persistent pursuit since the inception of the discipline, notably evident in Turing's seminal works predating the formalization of AI itself. Turing sought to establish the viability of thinking machines, laying the groundwork for subsequent reflections on attributing intelligence to artificial artifacts, particularly in the form of software programs. This chapter deals with the issue of measuring intelligence in artificial artifacts, emphasizing the importance of considering the expected intelligence. From an interactive standpoint, the human expectation of intelligence in an artificial entity revolves around performances deemed cognitively suitable for meaningful interactions. This perspective advocates for the detection of varying degrees of intelligence that align with the requirements of reliable and cognitively relevant interactions, even at intermediate levels. The discussion delves into the inherent differences between the simulative nature of human intelligence in artificial artifacts and the distinct characteristics expected for genuine intelligence in interactive contexts, particularly within the realm of social robotics. Indeed, the analysis of specific aspects of intelligence in robotic artifacts may prove enlightening by showing how the measurement of intelligence is significantly contingent upon the expectations that the human participant retains during interaction with AI. The inherently collaborative and interactive nature of the relationship between humans and robots underscores particular dimensions within the analysis of intelligence that might remain latent in interactions with disembodied AI systems.

**Keywords:** Measure of Intelligence, Expected Intelligence, Social Robotics.

# 1 Introduction

The issue of intelligence in artificial intelligence (AI) has been investigated since the dawn of the discipline, indeed it can be said that it was a foundational question. Turing's papers on machine intelligence addressed the problem before AI was even born [1]. In fact, Turing pursued the goal of establishing the possibility of machines, i.e. suitably programmed digital computers, which were capable of thinking. However, this starting point immediately led the British thinker to question the ways in which intelligence can be attributed to an artificial artifact, which in his reflection takes the form of a software program. Over the decades, the question has changed many times, developing in different directions that have followed the many evolutions of AI approaches [2]. In more recent times, the theme has also been re-proposed in the form of a possible measurement of intelligence in AIs, giving rise to further questions concerning the possibility of identifying, even quantitatively or along a scale, the presence of intelligence in artificial systems.

In this chapter we will address the problem of measuring intelligence in AIs arguing that it concerns and should consider more the expected intelligence of artificial artifacts. This is because from an interactive point of view, what the human being expects from an artificial artifact defined as intelligent is a set of performances that can be recognized as cognitively suitable for the interaction itself, so as to be recognizable and reliable. In this sense, we can speak of the detection of a, more or less high, degree of intelligence, which satisfies the requirements of a reliable and cognitively relevant interaction, even at intermediate levels. The investigation of general or human-level AI is certainly an interesting research field. We believe that the quantity of expected, and therefore attributable, intelligence, which artificial artifacts defined as intelligent must show, without this being a deception, but at the same time independently of the simulative nature of the human intelligence of the artifact itself, is equally important. We will therefore try to argue how the two aspects are only partially overlapping and we will provide some examples taken from the world of robotics, which, due to its nature as a discipline aimed at building situated and interactive entities with human beings, is suitable enough for such a kind of investigation.

The structure of the chapter is as follows. We will first analyze some aspects of the notion of measuring intelligence in natural and artificial systems (§ 2). We will then discuss the relationship between actual and expected intelligence in AI, also providing a proposal for parameterizable characteristics to measure expected intelligence (§ 3). We will then review some aspects of the development of AI to see its shortcomings and how this influences expectations (§ 4) and then move the discussion on robotics to interactive contexts (§ 5). Finally, after discussing some attempts to measure robotic intelligence (§ 6), we will draw some conclusions underlining possible future prospects for investigation (§ 7).

## 2 Measuring intelligence in natural and artificial systems

Unexpectedly, the issue of measuring intelligence has not been a dominant topic in AI from the start, as it has been the case with regard to the development of methodologies to measure human intelligence since the first steps in the early Twentieth century by psychologists such as Alfred Binet and Théodore Simon [3]. The issue of intelligence in AI was instead immediately addressed according to a very different scheme, which admitted only two possibilities: the presence or absence of intelligence [1]. The presence of intelligence, according to Turing, was what could be detected and attributed from the outside to a certain type of machines (suitably programmed discrete-state machines) in relation to their behavior. For this reason, the Turing Test in the following decades took the form of a yes/no test on the presence of intelligence in machines, i.e. programs which, in view of this possibility, fall within the vast and multifaceted field of AI (from software agents, to NLP applications, to problem solving algorithms, to robotics, and to neural networks, just to mention some of the most important AI areas). If this already marks the difference with human beings and the very goal of measuring their intelligence, the enterprise has proved to be even more difficult and somehow opaque as over the years there has been no effective definition of intelligence to be able to approach the intelligence of AI. The two problems seem to go hand in hand.

One of the aspects that has been underlined by those who have addressed this issue has been the difference between the types of intelligence involved: on the one hand, intelligence as a set of task-specific skills and, on the other, intelligence as a general ability of learning and open-endedly performing. This difference has been underlined for example by Chollet [4] in outlining his proposal for the measurement of AI intelligence. In fact, an AI focusing on a specific aspect and implementing a performance related to a particular skill seems more easily measurable in terms of intelligence, as it will eventually correspond to the degree of accuracy with which that performance is carried out. The measure of the success with which the system performs its task becomes a good marker for measuring the intelligence of the system, provided that the system actually performs the task for which it is designed in the way in which this is presupposed by the goals of the system itself, or "if our measure of performance captures exactly what we expect of the system" ([4]: 9). It's not hard to think of many AI systems where this type of measurement can be accomplished. The problem is different when trying to measure an AI system's ability to generalize, both in terms of a system's ability to deal with situations it has not previously encountered (a narrow form of generalization) and in relation to the ability to deal with situations that not even the developer of the system, as well as the system itself, has previously encountered (a broader form of generalization).

This aspect has also been present in the AI field since its dawn, for example in the attempt to generalize the abilities of the Logic Theorist into a General Problem Solver equipped with problem solving techniques that could be adapted to different types of problems and therefore independently of a specific type of content (in the case of the Logic Theorist, issues of logical scope and content). This declared goal by the authors of the program [5] later proved to be much harder to achieve than initially believed due to various problems connected with the possibility of operating within not always totally defined domains (unlike totally defined ones, the so-called toy domains) as well as with the adequacy of the represented and available knowledge to the program, just to mention the most relevant obstacles. The search for the generality of AI, yet, was not only a problem, but also a driving pressure in the evolution of AI approaches, leading to numerous results also theoretically. Among them we can mention, as an almost direct consequence, the approach of cognitive architectures such as SOAR, ACT-R and others, which constituted one of the main attempts to build AI systems provided with a greater power of operational generalization [6].

Another important aspect from this point of view is that the generalization of AI systems, in many approaches and, in particular, in cognitive architecture approach, has linked the possibility of achieving this kind of generalization to the reconstruction of cognitive abilities in the broad sense, and therefore to a cognitive matrix. This could lead us to consider the problem of measuring intelligence in an AI system of cognitive inspiration more easily solvable, precisely because the same techniques that are used to measure human intelligence seem relatively equally easily applicable, at least in principle, to cognitive artificial systems, i.e. to cognitive AI. If on one hand this may be true, on the other it should not be forgotten that a large part of the AI systems that today show a performance considered as very intelligent are machine learning systems, which therefore include a specific reference to learning, based on neural networks also of the deep kind. These systems, however, have given rise to a further explanatory problem. In many cases, they operate as a black box and it is not possible to recognize the reasons why they achieve very high results in terms of precision and accuracy, due exactly to the nature of the system itself, which is therefore not "explainable". This has led to the development of a continuously updated field of techniques to overcome this problem, that of Explainable AI or XAI [7]. These systems seem to inherently escape a measurement of their level of intelligence, yet, at least in a cognitive sense of the term. What can be measured in these AI systems, which can largely be traced back to decision-making performance in terms of proposed outputs, is the accuracy of the results compared to expectations. Thus, also in this case, what is measured is performance, without having any possibility of defining whether it is intelligent or not, i.e. the result of an activity that can be qualified as intelligent. In more abstract terms, it can be said that these algorithms are highly generalizable - a same technique that can be applied to many domains depending on the starting dataset - but this does make them opaque to an explanation and measurement of their intelligence.

An extensive analysis of the different attempts to measure the intelligence of AI systems shows that the ways of measuring (which usually boils down to a performance evaluation) of an AI system are many, but also very fragmented and specific for different systems, limiting its overall success. It seems difficult to obtain a reference standard for measuring intelligence that can be used for all AI systems. Furthermore, the distinction between a task-oriented assessment and a cognitive skills assessment, which is proposed by Hernádez-Orallo [8], seems to recall the aforementioned distinction between task-specific skills and intelligence as a general ability of learning and open-endedly performing. In this case, however, the distinction is motivated by a greater evaluation capacity of implemented methods and metrics. For example, in the case of assessing systems built to carry out a specific task, what is evaluated is performance rather than intelligence. Consequently, there is a risk that more than the AI system, those who designed and built/programmed it will be evaluated. This is certainly a useful evaluation, even if it is far from being a measurement of intelligence in AI. Hernádez-Orallo identifies three types of task-oriented evaluation: human discrimination, problem benchmarks, peer comparison. Without going too far into the details of each of these kinds of methods, we can draw attention to the fact that while the first is rather subjective, the others aim to be more objective by posing, in different ways, a parameter or several reference parameters, whether they are a standard or a result of a comparison (e.g. a match, a game, a competition).

An ability-oriented evaluation can lead to a better measurement of intelligence, even in progressive terms; for example, in relation to how much a system that is based on continuous updating due to machine learning techniques develops a progress in its intelligence over time. The reference to the measurement of abilities involves more general cognitive aspects such as reasoning skills, inductive learning abilities, motion abilities, etc. [8]. Even in this case, however, the problem oscillates between a greater anthropocentrism, such as for example with the use of psychometric tests, also those that are adapted to AI as in the case of Psychometric AI [9]; and, on the other hand, a stronger search for an objective and neutral quantification, with respect to the human capabilities, of the system's abilities, for example through the use of metrics derived from the algorithmic information theory, which is based, among other things, on the Kolmogorov notion of computational complexity [10]. The proposal of a universal psychometrics [11] seems to be a way to overcome the impasse due to the dichotomy between anthropocentric methods and the search for formal standards (encompassing general and neutral tests) for measuring performance in the case of general cognitive abilities. Nevertheless, even in this case it does not seem possible to completely escape a sort of anthropocentrism which, as regards general-purpose systems, moves towards the notion of cognition and the definition of cognitive ability. This becomes even more evident if we systematically consider the methods for measuring intelligence, combining those for natural intelligence and those for artificial intelligence [12], despite the many developments at least in terms of discussion on such a topic that have been accomplished in recent years [13]. Somehow, it seems impossible to measure intelligence without attributing such general ability, or the vehicles of that ability, to something external to the system (a performance in a specific task or a more general skill) and against an external standard to the system, but belonging to other intelligent systems (basically the human beings).

# 3 On actual and expected artificial intelligence

This view related to the evaluation and measurement of intelligence in AI systems is clearly problematic and it also seems hard to see how it is possible to go in a direction that constitutes an advance with respect to the impasse between specific skills evaluation and general ability evaluation. Such an impasse seems to give support to those who propose a deep transformation of our expectations on AI [14]. According to this view, artificial intelligent systems do not have much intelligence, indeed they probably do not have any, at least in the human sense of the term. They are machines, of course, and they are useful machines, however not for their intelligence, but as tools that are capable of producing good or excellent results. In this way, AI becomes a set of tools acting in a surprising way to achieve the expected results (excellent execution of the task). As an ability to act, AI systems have their strength in being interactive, autonomous and in the capability to learn and self-learn.

According to this view, AI exists, but not as intelligence per se. It is a very deflationary, perhaps radical, framework of AI, but perhaps less implausible than the catastrophic standpoints connected to the most recent developments of AI. On the other hand, if it is reasonable to consider such a framework as radical, it is also true that *an intelligent result is not always an intelligent behavior, or the result of an intelligent behavior.* This discrepancy, which is at the basis of the dissociative vision between intelligence and action of AI systems, now seems to be no longer contestable. The most recent AI developments show that the engineering solutions to achieve the most accurate performances are also those that lead to the best systems to replace humans in carrying out tasks considered intelligent or requiring human intelligence, without the use of such intelligence. In this case too an impasse is created, which goes beyond the one generated by the impossibility of explaining how certain results are achieved by the system. It is the impasse one faces with respect to the concept of intelligence itself, when it is replaced by something that one fails to consider, in a fine-grained analysis or from a structural point of view, intelligent.

However, there is a perspective that can be considered complementary, which sees AI systems as artifacts or social machines endowed with the peculiar features of being autonomous, self-learning and capable of carrying out a teleological, goal-driven behavior in their social interaction [15]. The social aspect makes the goal-oriented nature of these machines relevant in general, since their purpose is not that of those who designed them, nor that of the users with whom they interact, but the one which arises from the interaction between system and user. If considered broadly, this interpretative framework embraces a multiplicity of AI systems, from software managing the performance of many online platforms, to search engines, to interactive humanoid and non-humanoid robotic systems, just to name a few. We argue that in all these cases, and therefore in most cases in which human users are involved with AI systems (almost everyone when AI is involved), *it is essential the attribution of intelligence the human user does, often in real-time, to obtain good results and an interactive behavior as optimal as possible.* In other words, attributing intelligence on the part of human users seems an unavoidable requirement for the correct epistemic, applied and ethical functioning of these systems. From an epistemic standpoint, the attribution of intelligence can be seen as a cognitive necessity for users attempting to comprehend the system's responses and actions. It serves as a cognitive shorthand, allowing individuals to navigate the complexity of AI interactions by ascribing human-like understanding and intentionality to the machine. In this way, users can better anticipate and interpret the system's behavior, facilitating a smoother and more intuitive collaboration. On the other hand, applied functionalities of AI systems, especially in decision-making processes, benefit significantly from the user's attribution of intelligence. Furthermore from the ethical point of view, recognizing the user's role in attributing intelligence underscores the responsibility of designers and developers to create AI systems that align with human values and ethical standards. In other words, the attribution of intelligence serves as a crucial bridge between the designed purpose of the system and the dynamic, evolving nature of human interaction. Recognizing and attributing intelligence to AI systems fosters a symbiotic relationship where the machine's goal-oriented behavior aligns with user expectations and needs.

At this point one may wonder how it is possible to characterize this notion of intelligence attribution. Of course, there have been many well-known attempts, from Turing onwards, to define the attribution of intelligence to artificial systems, and, from Dennett [16] onwards, to define the attribution of intentionality. However, we would like to propose a pragmatic method for attributing intelligence, with the awareness that it also has complex theoretical and ethical implications. We therefore propose the following characteristics that an artificial intelligent system should/could have according to the user's perspective:

1. intelligence is attributed to the system before its use or interaction with it (on the basis of the preliminary knowledge that the user has of it);

2. intelligence is attributed to the system during its use or interaction with it;

3. the qualification of intelligent is attributed to the results of the system performance or of the interaction with it;

4. the repeated uses or repeated interactions with the system over time allow an attribution of intelligence subject to time variability.

If each of these characteristics is parameterized using a scale of values we can obtain an overall measurement of the users' expected intelligence related to the AI system they are using or with which they are interacting. This measurement can vary over time (especially in consideration of the parameter derived from item 4) and can also be refined with more detailed scales of values, thus producing a finer-grained differentiation, depending on the aims to be achieved. The overall goal remains to create a tool that is as flexible as possible for attributing intelligence to AI systems. The limit that can be recognized in this tool is that of being user-dependent. However, this is also one of its strengths, when such a tool is used to help draw up policies for using or engaging with AI systems, an increasingly pressing problem from a political, social and normative point of view. In addition, it has further strengths. For example, it is not a tool that focuses on the idea of purpose or goal-oriented behavior, however not excluding this characteristic from the evaluated AI system.

Furthermore, from the point of view of aggregated data (of individual user evaluations) it can allow an overall measurement of the intelligence expected by AI systems that allows an assessment not only in terms of policies, but also related to the general notion of intelligence, which could be reconstructed starting from the aggregated data of many users by making hypotheses on the various motivations that lead to certain attributions of intelligence. Our assumption is that expected intelligence (on the part of human being) is coupled with something in the system allowing intelligent behavior in general. This does not imply that such a thing is the same as in human beings, but at least is something that gives rise to intelligence, otherwise everything could potentially be described as, or ascribed with, intelligence without any particular reason. The reason (a mechanism, a technique, a dynamical interaction, a mathematical/statistical function or whatsoever) should be, to not lose the general notion of "artificial intelligence". In fact, this is not the case in the real world we live.

Finally, the user would in any case be safeguarded in his interaction with AI systems by a more adequate and at the same time transparent use of an artifact which calls into question the explicit awareness, on the part of the user, of the fact that he is interacting with AI systems equipped of different recognizable or explicit levels of intelligence.

A proposal of this kind is only preliminary and certainly needs more in-depth investigations. In principle, however, it can be a first step towards overcoming the impasse on the actual evaluation of the intelligence of an artificial system, which allows to consider intelligent artificial artifacts that are classifiable as such, without committing oneself to the metaphysical, epistemological or mechanistic reasons of such intelligence and, at the same time, without completely discarding the notion of intelligence in analyzing these systems. Furthermore, the more complex theoretical-epistemological investigations could in any case be carried out ex post, downstream of this type of measurement.

The analysis of specific aspects of intelligence in robotic artifacts may prove enlightening by showing how the measurement of intelligence is significantly contingent upon the expectations that the human participant retains during interaction with AI. Indeed, the inherently collaborative and interactive nature of the relationship between humans and robots underscores particular dimensions within the analysis of intelligence that might remain latent in interactions with disembodied AI systems. In the following sections, we will try to show that the issue of human expectations towards artificially intelligent systems explicitly emerged from contemplation on what first examples of AI lacked and how the demand for a collaborative, human-like form of intelligence gave rise, in addition to philosophical and technical inquiries, to novel modes of intelligence assessment and evaluation.

## 4 What is lacking in artificial intelligence and how it affects expectations

The initial years of AI were marked by a prevalent notion that human intelligence could, in principle, be comprehended and replicated in a machine [17]. This idea was first conceived during the notable gathering held at Dartmouth in 1956 [18]. At the time, intelligence was predominantly perceived as the ability to process data through computation, involving the deliberate application of suitable inferences for a given purpose [1][19]. Consequently, the field of AI has primarily focused on enhancing the logical and inferential capabilities of machines, with the belief that machines would achieve human-like results once their computational capacity reached a level of complexity equivalent to human intelligence.

Despite a clear agenda and initial successful applications, the field of AI encountered increasingly complex challenges, eventually entering a period of setbacks known as the AI Winter [20]. During the late 60s, several factors contributed to the negative impact on this research domain. Firstly, the recognition of the complexity of many problems posed difficulties as the frameworks proposed by Marvin Minsky oversimplified the issues, resulting in early systems excelling only in simplistic tasks. Real-world problems proved to be too intricate to solve, and scaling up the capabilities of AI systems went beyond the realm of faster hardware and memory [21]. In addition, the "thinking humanly" approach that characterized early AI endeavors proved inadequate for solving the problems at hand. This approach, known as the symbolic approach [22], involved attempting to replicate human problem-solving methods without breaking down the main problem into possible solutions and formulating an algorithm. Moreover, in many cases, positive results have been prone to over-interpretation, leading the public to believe that AI systems possess much greater intellectual capabilities than they actually did. For instance, the renowned program Eliza, developed by Weizenbaum [23], was often perceived by participants as having achieved a genuine understanding of human problems, akin to that of a psychotherapist. However, the program lacked any true knowledge or comprehension of its interlocutors or their issues. The enthusiasm surrounding Eliza can be attributed to the significantly low expectations people had for AI systems, beyond purely logical performance, which allowed for quick excitement to arise.

During this period, the debate surrounding the reasons why intelligent machines failed to achieve the expected accuracy in tasks envisioned by computer scientists was primarily divided into two positions. On one hand, Minsky attributed the mistakes to a naive preconception held by computer scientists and AI practitioners regarding the nature of the mind and its various aspects. In a seminal paper, Minsky [24] addressed the prevalent questions of that time, which continue to be relevant today, regarding the capabilities (or limitations) of intelligent machines. Characteristics such as creativity, non-logical thought, and self-awareness were deemed by computer scientists, brain researchers, and cognitive scientists to be inherently human and beyond the realm of implementation in machines. This notion profoundly influenced the development of AI. The expectations for machine performance were confined to logical tasks and the successful execution of programmed instructions, devoid of any consideration for creative problem-solving or deviations from what the programmer had explicitly taught the machine. However, Minsky argued that "all those beliefs which set machine intelligence forever far beneath our own are only careless speculations, based on unsupported guesses on how human minds might work" ([24]: 15). He suggested that it was necessary to redefine our conception of intelligence, not only with regards to machines but also in relation to ourselves as humans, and to place greater trust in the power of our intuition, which had been instrumental in constructing the initial AI models.

On the opposing side of the debate, in stark contrast to Minsky's stance, certain philosophers with alternative perspectives on mind and cognition expressed skepticism regarding the potential achievements of AI as a whole. This skepticism stemmed from ontological disparities between humans and machines and the fact that computer scientists embarked on their work without any knowledge of the philosophical attempts to address the same problems. Drawing on the phenomenology of Heidegger and Merleau-Ponty, as well as Gestalt psychology, Dreyfus argues that "what distinguishes persons from machines [...] is not a detached, universal, immaterial soul but an involved, self-moving, material body" ([25]: 149). The core argument posits that the problems confronted by computer scientists had already been contemplated by philosophers, and the arguments developed by the latter could be applied to the discourse on meaning and AI. The problem of meaning is a longstanding one in philosophy, and the manner in which computer scientists approached it traversed the history of philosophical thought on the subject, from Plato onwards. Dreyfus anticipated the failure of proposed plans for general problem solvers and automatic translation machines due to computer scientists' naive conception of mental functioning. As a result, he recommended that researchers familiarize themselves with modern philosophical approaches to human beings and intelligence if they aimed to replicate the characteristic aspects of human intelligence.

It is possible to argue that the positions put forth by Minsky and Dreyfus originated from the same recognition of the naivety of computer scientists, yet they were approached in divergent manners, resulting in distinct solutions and future agendas. In essence, Dreyfus believed that these limitations were practically insurmountable, starting from the phenomenological perspective on the body and consciousness. Conversely, Minsky advocated for a revaluation of the concept of intelligence and maintained that the desired outcome could be achieved with appropriate adjustments. However, Minsky's proposals proved to be more problematic than anticipated and sparked a new wave of contemplation on the feasibility of implementing mechanisms of signification and constructing machines that genuinely exhibit concern for their actions, as exemplified by the frame problem [26] and the symbol grounding problem [27].

Starting from these problems and recognizing the significant role of the body in the emergence of intelligent behavior, Brooks initiated a revolution in the field of AI and robotics, giving rise to a new paradigm known as "nouvelle AI" [28]. Indeed, in the 1980s, Brooks charted a course for redefining the approach to building artificial machines that could exhibit human-like intelligence and behavior through embodiment, which had profound implications for the subsequent development of ideas in cognitive sciences and philosophy of mind [29]. His proposal emphasized the necessity of constructing complete agents that operate in dynamic environments using real sensors in order to truly test the concepts of intelligence [28]. This required moving beyond the symbolic artificial intelligence model that had characterized previous approaches. Consequently, a new framework emerged, in which intelligence was no longer viewed solely in terms of accomplishment of isolated tasks but primarily in terms of interaction with the environment. The central idea was that "the true details of interacting with the world are not the same as abstract thinking has led many workers in Artificial Intelligence to believe" ([30]: 1). Brooks' work significantly propelled the field of robotics, which had previously progressed at a slower pace compared to disembodied intelligent systems. This breakthrough yielded numerous advancements, leading to the development of robots capable of interacting with both the environment and humans in a social manner, such as Kismet [31]. The most notable transformation brought about by this new paradigm in AI was the abandonment of the "human thinking" approach in favor of a perspective in which intelligence arises from more fundamental processes involving the interaction between an organism's body and its environment.

This reconfiguration of the concept of intelligence represents a significant model shift: intelligence shifts from being perceived as demonstrable solely through the correct execution of tasks to a conception wherein the foundational aspect of humanoid intelligence lies in how an agent interacts with its world. Most importantly, the work on robotics emphasized that the pursuit of disembodied tasks and the singular focus on developing computational capabilities, detached from immersion in an environment, were insufficient for achieving AI on par with human intelligence. However, the shift in focus from task performance to interaction with the world has presented challenges in evaluation, particularly when human agents are involved. Indeed, in the context of interactive AI, especially embodied AI over the past three decades, evaluation metrics have become more nuanced and now consider elements that were previously deemed irrelevant. As highlighted by Minsky, while the expectations of AI performance within the specialized field of computer science influenced internal considerations, *the desire to deploy robots in real-world settings and have them interact with human users without prior knowledge of their inner workings has brought the issue of expectations to the forefront.* Regardless of the level of accuracy with which robots can solve various tasks, even surpassing human capabilities, the perceived level of intelligence in robots depends on multiple variables, often rooted in embodied communication and extending beyond the mere explicit accomplishment of the task at hand.

## 5 Expected intelligence for robots in interactive contexts

One of the main factors that has been demonstrated to influence human expectations in interactions with robots is the degree of human-likeness in the robot's appearance. Extensive research has shown that when robots exhibit a high degree of human-likeness, it elicits anthropomorphic tendencies in humans, leading them to engage with the robots in a manner that resembles social interactions with other humans [32][33]. However, this anthropomorphic tendency can be seen as a double-edged sword. While it promotes social engagement, it also creates a potential for misleading expectations regarding the cognitive capabilities of the robots.

In the field of Human-Robot Interaction (HRI), this challenge of understanding the capabilities of robotic agents in interactive contexts has been identified as the Perceptual Belief Problem (PBP). As highlighted by Thellman and Ziemke [34], the PBP appears to be unique to the field of HRI and does not arise in standard Human-Computer Interaction. Unlike computers, social robots have physical bodies that enable them to navigate and engage with a complex and unstructured physical world, and at the same time bodies make robots able to physically interact with humans. The PBP in HRI arises because humans, when interacting with social robots, need to assess the robots' abilities and limitations based on their perceptual and cognitive capabilities. Since social robots often exhibit human-like appearances and behaviors, humans may attribute human-like cognitive capacities to them, creating potential misconceptions about the robots' actual cognitive abilities. This anthropomorphic tendency can lead to the *expectation that the robot possesses comprehensive perceptual beliefs similar to those of humans*. However, fulfilling such expectations is challenging given the current technology and robot architectures. Although a humanoid robot may have human-like eyes, ears, arms, and legs, it does not necessarily imply that it can see, hear, grasp objects, shake hands, or walk side by side in the same way as a human would in interaction. Thus, when confronted with a humanoid robot, human agents may be disappointed to realize that the robot's capabilities are limited. Moreover, *these expectations are difficult to measure* because assumptions about the actual cognitive and behavioral capabilities of robots are not always transparent to individuals. For instance, even if people treat the robot as an intentional agent during interaction, when asked whether they believe the robot has a mind, they typically respond in the negative [35]. Consequently, new methodologies have been developed to evaluate the PBP without relying solely on verbal self-assessment [36]. Finding a way to address the gap posed by the PBP is crucial to ensure a high degree of collaboration and *to prevent disappointment in humans*.

A high degree of human likeness can also have negative effects on interaction when combined with a low level of affinity toward the robot. This phenomenon is known as the Uncanny Valley Effect [37], wherein a negative feeling arises in human agents interacting with robots that closely resemble humans but lack the same level of human-like behavior. Various explanations have been proposed to account for this effect, but consensus has not yet been reached. One of the most widely accepted explanations [38] attributes the effect to a mismatch between human agents' expectations, influenced by the humanoid appearance, and the actual behavior exhibited by the robot in interaction. Recently, this explanation has also been examined through the lens of the theory of predictive coding [39][40], which posits that the fundamental functional structure of the brain, across all levels of organization, involves the comparison of observations with predictions and strives to operate in a way that minimizes any discrepancies between them.

Not only do humanlikeness and anthropomorphism play a crucial role in shaping expectations regarding robot intelligence in interaction, but the cultural background and representations of robots in cultural settings also significantly impact human expectations. As Kamide and Mori [41] emphasize, culture and philosophy have a profound influence in the context of Human-Robot Interaction (HRI). In their study, they compare the philosophical systems of the West (e.g., Europe and the Americas) to those of the East (e.g., Asia and the Middle East). The Western tradition seeks a systematic, consistent, and comprehensive understanding of the universe, while the Eastern tradition adopts a more holistic or circular view of the world. These differences in philosophical orientations can lead to varying attitudes towards robots, suggesting that cultural and philosophical leanings may foster greater readiness for acceptance among Eastern populations, as also suggested by MacDorman et al. [42] in the context of Japanese culture. Indeed, cultural narratives in which individuals engage from childhood significantly influence the acceptance of robots in society. In Japanese culture, for example, where robots have been portrayed as companions and allies in manga adventures since the 70s, the majority of people tend to exhibit a positive outlook towards the integration of robots engaging in daily activities in close proximity to the general public [43]. Conversely, individuals from Western cultures often harbor more negative attitudes regarding the inclusion of robotic companions for assistive and care tasks, influenced by apocalyptic narratives found in cinema and literature, such as the works of Asimov or the Terminator movies [44][45]. On the same topic, Horstmann and Krämer [46] conducted a study revealing that individuals' exposure to media portrayals of social robots significantly influences their expectations of robots' abilities and capabilities, subsequently reinforcing and amplifying those expectations. Furthermore, the study found that individuals' knowledge of negatively depicted fictional social robots contributes to the development of negative expectations, perceiving robots as potential threats. Conversely, individuals who possess a greater understanding of the capacities and limitations of robot technology, based on non-fictional knowledge, exhibit reduced levels of anxiety towards robots. In a follow-up study [47], they examined how a negative violation of expectations caused by a social robot, coupled with the valence of the subsequent reward, could influence participants' desirability of interacting with a social robot and its impact on HRI. Interestingly, the study revealed that when the robot violated participants' expectations, they evaluated the robot's competence, sociability, and interaction skills more negatively [1].

Therefore, the framework for evaluating robotic intelligence needs to account for these expectations, both prior to actual interactions and during online interactions with robots. In the following paragraph, we will explore how incorporating these new insights can further enhance the understanding and assessment of expectations in HRI. Although no study, to our knowledge, specifically investigates the role of these expectations in influencing the evaluation of robot intelligence, certain tests have been developed to measure perceived intelligence in robot interactions, along with related metrics that facilitate the overall assessment of HRI quality.

## 6 Measuring robotic intelligence

In the interactive context, the measurement of a robot's intelligence is an assessment of how effectively a human and a robot collaborate [50]. As discussed in the previous paragraph, human expectations regarding intelligence are not only relevant in terms of preconceived beliefs and assumptions, but also during the actual interaction with the robot. Negative attitudes, in particular, can hinder the potential benefits of HRI, such as therapeutic practices.

To evaluate these assumptions and attitudes towards robots, Nomura et al.

---

[1]The influence of cultural background on expectations towards robots is an extensively explored aspect in the Human-Robot Interaction literature. Delving deeper into this topic is beyond the scope of the present chapter; see [48; 49] for a general overview.

[51] developed the Negative Attitude towards Robots Scale (NARS). This scale consists of a questionnaire that explores various aspects of human-robot interaction, including situations, social influence, and emotions. In their experimental study, participants engaged with Robomovie, a robot designed for human communication [52], and they were asked to interact verbally with the robot and to touch it. The findings revealed that negative attitudes towards robots can indeed impact the interaction. Some individuals displayed resistance to touching or speaking with the robot, which undermined and compromised the quality of interaction. Additionally, the study showed that these attitudes can vary depending on the participants' gender and culture. Furthermore, previous encounters with robots were found to influence the results of the NARS test. Participants with prior direct experience with robots were more inclined to engage in verbal interactions and touch the robot during the study. Subsequent studies examined the efficacy of the NARS scale in assessing changes in expectations during actual interactions [53] and explored cultural differences among participants [54].

Other assessment methods, often used in conjunction with the NARS test, have been developed to evaluate the overall quality of HRI after the initial validation stage. One widely employed approach is the Godspeed test [55], which aims to ensure comparability across different experimental settings and the replicability of results in various languages and contexts. This test encompasses several factors, including anthropomorphism, animacy, likeability, perceived safety, and perceived intelligence, all of which are assessed based on participants' experiences during the interaction. Of particular relevance to our study is the assessment of perceived intelligence. The researchers conceptualize perceived intelligence in robots as a combination of robotic competence and the potential elicitation of the perception of intelligence through "random behavior" during interaction. Participants were asked to rate their perceived intelligence on a scale ranging from 1 to 5 of different dichotomies, i.e., Incompetent/Competent, Ignorant/Knowledgeable, Irresponsible/Responsible, Unintelligent/Intelligent, and Foolish/Sensible. Although the test has undergone statistical validation, it is important to highlight two key considerations.

Firstly, attempting to capture perceived intelligence through strict dichotomies, as described above, may not provide a comprehensive assessment of quality. Regarding perceived intelligence, certain terms in the Godspeed test appear to be redundant (Incompetent, Ignorant, Unintelligent) and may not be easily discernible. Other terms (Irresponsible/Responsible, Foolish/Sensible) seem unrelated to the evaluation of perceived intelligence but are more relevant to ethical considerations. Dichotomies such as Intelligent/Unintelligent and Ignorant/Knowledgeable align better with the general concept of perceived intelligence, as they reflect stereotypical conceptions of intelligence. This makes us think that this list can be restricted or enlarged at will, also because the authors did not give any explanation for having chosen these terms and no others. The second issue to emphasize is that the Godspeed test alone may not adequately capture the overall quality of HRI. As previously mentioned, the embodied factor plays a crucial role in robot interactions and is not considered in this test. Therefore, the integration of implicit measures of perceived intelligence, which assess the bodily attitudes people exhibit during robot interactions, could enhance the evaluation process. Such measures have been adopted in various experimental settings [56][57] and would contribute to a more comprehensive framework for assessing interactions while also enhancing internal test validity.

In relation to the impact of expectations on human-robot interaction, Rosén et al. [58] developed an evaluation framework aimed at providing a comprehensive understanding of the social robot expectation gap. This term describes a disparity in which expectations, whether excessively high or too low, can result in disconfirmation. The authors illustrate this phenomenon by considering two scenarios: firstly, when individuals interact with a social robot and hold the belief, influenced by depictions in science fiction movies, that the robot could feel and express pain when it falls, yet it does not, leading to a contradiction of expectations and the emergence of high expectations; secondly, when users do not assume the robot to be able to engage in verbal communication, but it initiates a conversation, leading to a contradiction of expectations and the development of low expectations. Consequently, for the authors it is possible to say that the quality of interaction can be perceived as high, irrespective of the robot's capabilities, as long as expectations are confirmed. To assess the implications of high and low expectations on social robots prior to, during, and after interaction, the authors proposed an evaluation framework encompassing affect, cognitive processing, and behavior and expectations. Moreover, they highlighted that expectations are dynamic and change over time - in line with our previous proposal - but they do not provide experimental evidence for the effectiveness of this evaluation model.

However, it is crucial to develop a means of evaluating robotic intelligence in order to facilitate suitable human-robot interactions and design effective robot behaviors. To gain a comprehensive understanding of the extent to which robots can exhibit intelligence, Winfield [59] proposed a framework in which robotic intelligence arises from the integration and interaction of four distinct types: morphological intelligence, swarm intelligence, individual intelligence, and social intelligence. Using a star diagram, the author compared different organisms based on their exhibited intelligence. For example, humans may demonstrate high levels of individual and social intelligence, surpassing other animal species, but exhibit a lower degree of swarm intelligence, in contrast to certain animal species like ants. This framework can also be applied to robots, allowing for an analysis of their diverse forms of intelligence. Winfield further suggested that one possible reason for human disappointment in terms of robotic intelligence is that "none of the intelligence graphs for the robots score on more than two axes, whereas all of the exemplar animals score on at least three" ([59]: 6). This highlights a distinction between the types of intelligence exhibited by living beings and those demonstrated by artificial artifacts.

Despite the fundamental difference that Whitfield's conclusions highlight between living beings and artificial artifacts, the disparity in terms of the intelligence exhibited by these entities appears to be once again focused on restricted aspects of intelligence. In the majority of cases, intelligence is conceptualized as a quality demonstrated through abstract reasoning for specific tasks and subsequently evaluated by comparing it to the best possible outcomes, which are ultimately determined by how humans accomplish such tasks.

These examples taken from HRI show how the problem of expected intelligence is widely present in interactive robotics, a field in which intelligence can be assessed on multiple dimensions including physical ones (performance in specific tasks, acting in the real world, verbal and non-verbal interactions with human users, etc.). Every test shows some of the characteristics that we have included in our general list, although not all at the same time. However, it should appear evident, by means of the HRI, as a general tool for assessing expected intelligence and for drawing up a shared and aware metric, is a theoretical and a practical problem that can no longer be postponed.

# 7 Conclusions

In conclusion, two final considerations can be made. The first pertains to the often inadequate consideration of expectations in most tests aimed at assessing or evaluating intelligence in artificial artifacts. These tests often fail to fully acknowledge the role of expectations held by both human users and scientists regarding their overall understanding of the interaction and the task at hand. While expectations may sometimes be examined from the perspectives of anthropomorphism or cultural background, the assessments rarely, if ever, take into account the variety of expectations that users may have when engaging with artificial artifacts. Furthermore, in the case of robots, apart from the factors we have previously discussed as influencing expectations, issues concerning the attribution of intentionality [57] and perceived agency [60] are also considered to be influential factors that affect the actual interaction with robots and consequently impact the evaluation of their intelligence.

At the same time, these attempts for assessing and evaluating intelligence strive to break it down into subsections, thereby seeking to simplify the problem. Indeed, mainstream methods separate different abilities and evaluate them individually, subsequently aggregating the singular evaluations. Consequently, the overall evaluation of the artificial artifact's intelligence is understood as the sum of the separate evaluations of its performances, whether accomplished or not. As emphasized by Anzalone and colleagues [61], the challenge in evaluating AI lies not in reducing it to the efficacy of individual algorithms, but in offering a more comprehensive account that encompasses broader forms of intelligence, namely social and interactive intelligence [62]. Therefore, what we propose is to complement this purely methodological evaluation approach with aspects derived directly from interaction studies, in order to facilitate a more robust evaluation that takes into consideration various factors that we have observed to contribute to the assessment of a robot's intelligence to varying extents, such as expectations and the overall quality of interaction.

Finally, if the conception of intelligence that we wish to apply to artificial artifacts is derived from or must be comparable to human intelligence, it is imperative to carefully consider the precise basis on which we are aligning them. The response to the question "how can an artificial artifact be deemed intelligent?" will always be influenced by the particular concept of intelligence that is embraced.

It is not possible to say whether a final answer will ever be given on what intelligence is and on what intelligence is from the point of view of the artificial. However, we can consider it reasonably certain that the attribution of intelligence by the human user is an unavoidable characteristic, both from a cognitive and a practical point of view. This also has positive motivations and implications, as we have tried to argue. The proposal we have made is that taking into account the intelligence expected in artificial artifacts and attempting to measure it through metrics based on the various aspects of user interaction or use can lead to better results in understanding the reality of the artificial and, at the same time, to better designs of artificial artifacts. And perhaps to a more well-structured answer to the question of intelligence in AI. The list of characteristics proposed to measure expected intelligence is in line, as we have tried to show, with the measurement attempts made in social robotics, a field that seems to exemplify more than others in AI the strong implications that the question on the intelligence brings with it, constituting an area of investigation whose results can also be extended to the interaction and use with other AI systems which are more distant from an embodied configuration.

# References

[1] Turing, A. M. (1950). Computing Machinery and Intelligence. *Mind*, 59, 433–460, (reprinted in J. Copeland (ed.), *The essential Turing*, Oxford University Press, 2004, pp. 441–464).

[2] Moor, J. H. (ed.) (2003), *The Turing Test. The Elusive Standard of Artificial Intelligence*, Dordrecht, Springer. https://doi.org/10.1007/978-94-010-0105-2

[3] Esping, A., Plucker, J.A. (2015), Alfred Binet and the Children of Paris, in S. Goldstein, D. Princiotta, J. A. Naglieri (eds.), *Handbook of Intelligence, Evolutionary Theory, Historical Perspective and Current Concepts*, Springer, pp. 153-161.

[4] Chollet, F. (2019). On the Measure of Intelligence. *arXiv*:1911.01547v2, https://doi.org/10.48550/arXiv.1911.01547

[5] Newell, A., Shaw, J.C., Simon, H.A. (1959). Report on a general problem-solving program. https://exhibits.stanford.edu/feigenbaum/catalog/sy501xd1313

[6] Lieto, A. (2021), Cognitive Design for Artificial Minds. London, UK: Routledge, Taylor & Francis.

[7] Saeed, W., Omlin, C. (2023), Explainable AI (XAI): A systematic meta-survey of current challenges and future opportunities, *Knowledge-Based Systems*, 263, 110273, https://doi.org/10.1016/j.knosys.2023.110273.

[8] Hernández-Orallo, J. (2017a), Evaluation in artificial intelligence: from task-oriented to ability-oriented measurement, *Artificial Intelligence Review*, 48, pp. 397–447 DOI 10.1007/s10462-016-9505-7

[9] Bringsjord, S, (2011), Psychometric artificial intelligence, *Journal of Experimental and Theoretical Artificial Intelligence*, 23, pp. 271–277.

[10] Li, M, Vitányi, P. (2008), *An introduction to Kolmogorov complexity and its applications*, Springer, New York.

[11] Hernández-Orallo, J., Dowe, D. L. (2010), Measuring universal intelligence: Towards an anytime intelligence test, *Artificial Intelligence*, 174, pp. 1508–1539.

[12] Hernández-Orallo, J, (2017b), *The Measure of all Minds. Evaluating Natural and Artificial Intelligence*, Cambridge University Press, New York.

[13] Hernández-Orallo, J., Loe, B.S., Cheke, L., Martínez-Plumed, F., Ó Éigeartaigh, S. (2021), General intelligence disentangled via a generality metric for natural and artificial intelligence. *Scientific Report*, 11, 22822. https://doi.org/10.1038/s41598-021-01997-7

[14] Floridi, L. (2023), *The Ethics of Artificial Intelligence. Principles, Challenges, and Opportunities*, Oxford University Press, Oxford.

[15] Cristianini, N., Scantamburlo, T. and Ladyman, J. (2023). The social turn of artificial intelligence. *AI & Society*, 38, pp. 89–96. https://doi.org/10.1007/s00146-021-01289-8

[16] Dennett, D. C. (1987). *The intentional stance*. Cambridge (MA): The MIT Press.

[17] Cordeschi, R. (2002). *The discovery of the Artificial. Behavior, Mind and Machines before and beyond Cybernetics*. Dordrecht/Boston/London: Kluwer Academic Publishers.

[18] Boden, M. (2006). *Mind as Machine: A History of Cognitive Science*. Cambridge (MA): Oxford University Press.

[19] Russell, S.J, and Norvig, P.R. (1995) *Artificial Intelligence: A Modern Approach*. London: Pearson.

[20] Toosi, A., Bottino, A. G., Saboury, B., Siegel, E., and Rahmim, A. (2021). A Brief History of AI: How to Prevent Another Winter (A Critical Review). *PET clinics*, 16(4), 449–469. https://doi.org/10.1016/j.cpet.2021.07.001

[21] Dennett, D. (1984). Cognitive wheels: the frame problem of AI. *Minds, Machines and Evolution*, 129-151.

[22] Newell, A. (1980) Physical symbol systems. *Cognitive Science*, 4, 135-183.

[23] Weizenbaum, J. (1966). ELIZA—a computer program for the study of natural language communication between man and machine. *Communications of the ACM*, 9(1), 36-45.

[24] Minsky, M. L. (1982). Why People Think Computers Can't. *AI Magazine*, 3(4), 3. https://doi.org/10.1609/aimag.v3i4.376

[25] Dreyfus, H. L. (1972). *What computers can't do: The limits of artificial intelligence*. New York: MIT Press.

[26] Shananan, M. (2004) The frame problem. The Stanford Encyclopedia of Philosophy (Spring 2016 Edition), Edward N. Zalta (ed.), URL <https://plato.stanford.edu/archives/spr2016/entries/frame-problem/>.

[27] Harnad, S. (1990) The Symbol Grounding Problem. *Physica D* 42: 335-346.

[28] Brooks, R. (1991) Intelligence without Reason. *Proceedings of the 12th international joint conference on Artificial intelligence* - Volume 1 (IJCAI'91). Morgan Kaufmann Publishers Inc., San Francisco, CA, USA, 569–595.

[29] Varela, F., Thompson, E., and Rosch, E. (1991). *The Embodied Mind.* Cambridge (MA): MIT Press.

[30] Brooks, R. (1989). The whole iguana. *Robotics science*, 432-456.

[31] Breazeal, C. (2002). *Designing Sociable Robots.* Cambridge (MA): MIT Press.

[32] Airenti, G. (2018). The development of anthropomorphism in interaction: Intersubjectivity, imagination, and theory of mind. *Frontiers in Psychology*, 9, 2136.

[33] Damiano, L., and Dumouchel, P. (2018). Anthropomorphism in human–robot co-evolution. *Frontiers in Psychology*, 9, 468.

[34] Thellman, S. and Ziemke, T. (2021). The Perceptual Belief Problem: Why Explainability Is a Tough Challenge in Social Robotics. *Trans. Hum.-Robot Interact.* 10, 3, Article 29 (July 2021), 15 pages. https://doi.org/10.1145/3461781

[35] Fussell, S., Kiesler, S., Setlock, L., and Yew, V. (2008). How people anthropomorphize robots. In Proceedings of the 3rd ACM/IEEE international conference on Human-robot interaction (HRI '08). Association for Computing Machinery, New York, NY, USA, 145–152. https://doi.org/10.1145/1349822.1349842

[36] Thellman, S. and Ziemke, T. (2020). Do You See what I See? Tracking the Perceptual Beliefs of Robots. iScience, Volume 23, Issue 10, 2020, 101625, https://doi.org/10.1016/j.isci.2020.101625

[37] Mori, M. (1970). Bukimi no tani [the uncanny valley]. *Energy*, 7, 33-35.

[38] Kätsyri, J., Förger, K., Mäkäräinen, M., and Takala, T. (2015). A review of empirical evidence on different uncanny valley hypotheses: support for perceptual mismatch as one road to the valley of eeriness. *Frontiers in Psychology*, 6, 390.

[39] Saygin, A. P., Chaminade, T., Ishiguro, H., Driver, J., and Frith, C. (2012). The thing that should not be: predictive coding and the uncanny valley in perceiving human and humanoid robot actions. *Social cognitive and affective neuroscience*, 7(4), 413-422.

[40] Urgen, B. A., Li, A. X., Berka, C., Kutas, M., Ishiguro, H., and Saygin, A. P. (2015). Predictive coding and the Uncanny Valley hypothesis: Evidence from electrical brain activity. *Cognition: a bridge between robotics and interaction*, 15-21.

[41] Kamide, H., and Mori, M. (2016) One being for two origins - a necessary awakening for the future of robotics. In 2016 *IEEE workshop on advanced robotics and its social impacts (ARSO)*. IEEE, Piscataway, NJ.

[42] MacDorman, K.F., Vasudevan, S.K., Ho, C-C. (2009). Does Japan really have robot mania? Comparing attitude by implicit and explicit measures. *AI & Society*, 23: 485-510.

[43] Han, J., Hyun, E., Kim, M., Cho, H., Kanda, T., and Nomura, T. (2009). The Cross-cultural Acceptance of Tutoring Robots with Augmented Reality Services. *J. Digit. Content Technol. its Appl.*, 3, 95-102.

[44] Haring, K.S., Mougenot, S., Ono, F., Watanabe, K. (2014). Cultural Differences in Perception and Attitude towards Robots. *International Journal of Affective Engineering*, 2014, Volume 13, Issue 3, Pages 149-157, https://doi.org/10.5057/ijae.13.149

[45] Dumouchel, P., and Damiano, L. (2017). *Living with robots*. Cambridge (MA): Harvard University Press.

[46] Horstmann, A. C., and Krämer, N. C. (2019). Great expectations? Relation of previous experiences with social robots in real life or in the media and expectancies based on qualitative and quantitative assessment. *Frontiers in psychology*, 10, 939.

[47] Horstmann, A. C., and Krämer, N. C. (2020). Expectations vs. actual behavior of a social robot: An experimental investigation of the effects of a social robot's interaction skill level and its expected future role on people's evaluations. *PloS one*, 15(8), e0238133.

[48] Papadopoulos, I., and Koulouglioti, C. (2018). The influence of culture on attitudes towards humanoid and animal-like robots: An Integrative Review. Journal of Nursing Scholarship, 50(6), 653-665.

[49] Lim, V., Rooksby, M., and Cross, E. S. (2021). Social robots on a global stage: establishing a role for culture during human–robot interaction. International Journal of Social Robotics, 13(6), 1307-1333.

[50] Crandall, J. W., and Goodrich, M. A. (2003). Measuring the intelligence of a robot and its interface. In *NIST's Performance Metrics for Intelligent Systems Workshop*, Arlington, VA, 2003.

[51] Nomura, T., Suzuki, T., Kanda, T., and Kato, K. (2006a). Measurement of negative attitudes toward robots. *Interaction Studies. Social Behaviour and Communication in Biological and Artificial Systems*, 7(3), 437-454.

[52] Ishiguro, H., Ono, T., Imai, M., and Kanda, T. (2003). Development of an interactive humanoid robot "Robovie"—an interdisciplinary approach. In *Robotics Research: The Tenth International Symposium* (pp. 179-191). Springer Berlin Heidelberg.

[53] Nomura, T., Kanda, T., Suzuki, T., and Kato, K. (2008). Prediction of human behavior in human–robot interaction using psychological scales for anxiety and negative attitudes toward robots. *IEEE transactions on robotics*, 24(2), 442-451.

[54] Bartneck, C., Nomura, T., Kanda, T., Suzuki, T., and Kato, K. (2005). Cultural Differences in Attitudes Towards Robots. Companions: Hard Problems and Open Challenges in Robot-Human Interaction, 1.

[55] Bartneck, C., Kulić, D., Croft, E., and Zoghbi, S. (2009). Measurement instruments for the anthropomorphism, animacy, likeability, perceived intelligence, and perceived safety of robots. *International journal of social robotics*, 1, 71-81.

[56] Sciutti, A., Bisio, A., Nori, F., Metta, G., Fadiga, L., and Sandini, G. (2013). Robots can be perceived as goal-oriented agents. *Interaction Studies*, 14(3), 329-350.

[57] Thellman, S., and Ziemke, T. (2019) The Intentional Stance Toward Robots: Conceptual and Methodological Considerations. *CogSci'19. Proceedings of the 41st Annual Conference of the Cognitive Science Society*, 1097–1103.

[58] Rosén, J., Lindblom, J., and Billing, E. (2022). The Social Robot Expectation Gap Evaluation Framework. In *International Conference on Human-Computer Interaction* (pp. 590-610). Cham: Springer International Publishing.

[59] Winfield, A. F. (2017). How intelligent is your intelligent robot?. *arXiv preprint arXiv*:1712.08878.

[60] van der Woerdt, S., and Haselager, P. (2019). When robots appear to have a mind: The human perception of machine agency and responsibility. *New Ideas in Psychology*, 54, 93-100.

[61] Anzalone, S. M., Boucenna, S., Ivaldi, S., and Chetouani, M. (2015). Evaluating the engagement with social robots. *International Journal of Social Robotics*, 7, 465-478.

[62] Barchard, K. A., Lapping-Carr, L., Westfall, R. S., Fink-Armold, A., Banisetty, S. B., and Feil-Seifer, D. (2020). Measuring the perceived social intelligence of robots. *ACM Transactions on Human-Robot Interaction (THRI)*, 9(4), 1-29.

# FRAMING BELIEFS INTO FRACTIONAL SEMANTICS FOR CLASSICAL LOGIC

MATTEO BIZZARRI
*Scuola Normale Superiore*
matteo.bizzarri@sns.it

---

**Abstract.** The purpose of this paper is to explore one of the possible evolutions of fractional semantics for classical logic, initially introduced in [12]. Fractional semantics is derived by focusing on the axiomatic structure of proofs expressed in Kleene's sequent system $GS4$ [9, 16].

In this contribution, we aim to investigate the potential construction of a dialogue between logic and the "world", considering sets of beliefs while maintaining crucial proof-theoretic properties. To achieve this, we begin with the sequent system $GS4$. A set of sentences $B = \{b_1, \ldots, b_n\}$ is introduced to the sequent system as a cluster of new axioms, representing the beliefs of a certain rational agent. Sets of beliefs are treated in the standard manner as described in [3] and [5].

We denote the system obtained this way as $GS4_B$, with its closure relation denoted as $\vdash_B$. The philosophical idea underpinning this system is that a rational agent typically regards their own beliefs as true. Consequently, the beliefs in $B$ are automatically considered true and assigned a value of 1. $GS4_B$ qualifies as a supraclassical logic since $\vdash \subseteq \vdash_B$. This system could prove beneficial for further developments concerning decision processes in artificial intelligence.

**Keywords:** Proof Theory, Logics of Knowledge and Belief, Supraclassical logics, Many Valued Logics .

## 1 Introduction

The goal of this contribution is to explore one of the potential applications of fractional semantics for classical logic, which was firstly introduced in [12]. Fractional semantics for classical logic is a multi-valued semantics that is based on pure proof-theoretic considerations, where truth values are rational numbers within the closed interval [0,1]. The main difference between classical Boolean interpretation and fractional semantics is the asymmetry between classical tautologies and contradictions. Fractional semantics can assign values that differ from 0 to both non-logical axioms and contradictions, and it can be viewed as a method of determining how close a given proposition is to being a tautology or a contradiction.

To enable a fractional interpretation of its formulas, fractional semantics requires an appropriate proof-theoretic framework, which is a decidable logic $\mathcal{L}$ that can be displayed in a sequent system **S** (or its variants) as stated in [14].

Fractional semantics is obtained by focusing on the axiomatic structure of proofs expressed in Kleene's one-side sequent system $GS4$ [9, 16]. The system has the following rules:

$$\frac{}{\vdash \Gamma, p, \bar{p}} \ (ax.)$$

$$\frac{\vdash \Gamma, A, B}{\vdash \Gamma, A \vee B} \ (\vee)$$

$$\frac{\vdash \Gamma, A \qquad \vdash \Gamma, B}{\vdash \Gamma, A \wedge B} \ (\wedge)$$

There is not a rule governing negation as it is inductively defined by different atomic formulas $p$ and $\bar{p}$, where $\bar{p}$ indicates the negation of $p$. The interpretation of a formula is the result of the ratio between the number of identity top-sequents $(\Delta, p, \bar{p})$ out of the total number of top-sequents occurring in any of its proofs. Weakening and contraction are absorbed by rules, while cut rule has the form:

$$\frac{\vdash \Gamma, p \qquad \vdash \bar{p}, \Delta}{\vdash \Gamma, \Delta} \ (cut)$$

In order to give a fractional interpretation a counterpart is needed, namely $\overline{GS4}$, that is the $GS4$ calculus maximally extended:

**Definition 1.1** ($\overline{GS4}$). $\overline{GS4}$ is defined as $GS4$ calculus maximally extended, adding the complementary axiom schema which enables the introduction of whatsoever consistent clause $\vdash \Delta$. The system $\overline{GS4}$ is deductively trivial, i.e. anything can be derived in the system.

**Definition 1.2** ($[\![\Gamma]\!]$). $[\![\Gamma]\!]$, for each multiset $\Gamma$, indicates the fractional semantics value of $\Gamma$.

In $\overline{GS4}$, for instance, $\vdash p \to (q \land p)$ translated in $GS4$ as $\vdash \overline{p} \lor (q \land p)$ is derivable:

$$\cfrac{\cfrac{\vdash \overline{p}, q \; ax. \qquad \vdash \overline{p}, p \; ax.}{\vdash \overline{p}, q \land p} (\land)}{\vdash \overline{p} \lor (q \land p)} (\lor)$$

In classical logic the value of this sequent would be 0, but in fractional semantics it is possible to assign a value based on the number of tautological clauses out of two leaves in total, i.e., $[\![\overline{p} \lor (q \land p)]\!] = 1/2 = 0.5$, because of the presence of the axiom $\vdash \overline{p}, p$. Stability guarantees that any other decomposition of this sequent will always return the value 0.5. It is possible to give a formal definition of top-sequents axioms:

**Definition 1.3** (Top-sequents axioms).

$top^1(\pi)$ : denotes the multiset of all and only $\pi$'s *top-sequents* introduced by an identity axiom;

$top^0(\pi)$ : denotes the multiset of all and only $\pi$'s *top-sequents* introduced by a complementary axiom, in other words, those axioms that are not tautological.

Any formula $A$ can be interpreted as the ratio between the number of identity top-sequents (sequents introduced by the standard axiom) out of the total number of top-sequents.

$$[\![A]\!] = \frac{top^1(\pi)}{top^1(\pi) + top^0(\pi)}$$

**Example 1.1.** Let's take two examples to clarify the process. Let's firstly decompose the expressions $A$ and $B$, respectively $\vdash (p \lor q) \land (p \lor \overline{p}) \land (\overline{r} \land \overline{t})$ and $\vdash (p \lor \overline{p}) \land q \land (r \lor t)$. Let's start with $A$:

$$\cfrac{\cfrac{\cfrac{\vdash p, q \; (\overline{ax.})}{\vdash p \lor q} (\lor) \qquad \cfrac{\vdash p, \overline{p} \; (ax.)}{\vdash p \lor \overline{p}} (\lor)}{\vdash (p \lor q) \land (p \lor \overline{p})} (\land) \qquad \cfrac{\vdash \overline{r} \; (\overline{ax.}) \qquad \vdash \overline{t} \; (\overline{ax.})}{\vdash (\overline{r} \land \overline{t})} (\land)}{\vdash (p \lor q) \land (p \lor \overline{p}) \land (\overline{r} \land \overline{t})} (\land)$$

This proof contains one identity axiom out of four top-sequents in total and this means that:

$$\llbracket (p \vee q) \wedge (p \vee \bar{p}) \wedge (\bar{r} \wedge \bar{t}) \rrbracket = \frac{1}{4} = 0.25$$

Let's consider $B$:

$$\dfrac{\dfrac{\dfrac{\overline{\vdash p, \bar{p}}\ (ax.)}{\vdash p \vee \bar{p}}\ (\vee) \quad \overline{\vdash q}\ (\overline{ax.})}{\vdash (p \vee \bar{p}) \wedge q}\ (\wedge) \quad \dfrac{\overline{\vdash r, \bar{r}}\ (ax.)}{\vdash r \vee \bar{r}}\ (\vee)}{\vdash (p \vee \bar{p}) \wedge q \wedge (r \vee \bar{r})}\ (\wedge)$$

This proof contains two identity axioms out of 3 top sequents in total and this means that:

$$\llbracket (p \vee q) \wedge (p \vee \bar{p}) \wedge (\bar{r} \wedge \bar{r}) \rrbracket = \frac{2}{3}$$

The ultimate value of the sequent remains consistent across all its various decompositions. Fractional semantics has the capability to track the count of tautologies among the total axioms regardless of the decomposition order.

It is possible to define the fractional semantics in terms of a multi-valued logics, made by this definition:

**Definition 1.4** (Top-sequents). Top-sequents represent the number of the leaves of the proof as defined in Definition 1.3 and $\llbracket \vee \Gamma \rrbracket$ represents the value assigned to the multiset $\Gamma$ where only $\vee$–applications appear.

$top^1(\pi)$ : let's call this $m$

$top^0(\pi)$ : let's call this $n$

$\llbracket \vee \Gamma \rrbracket$ is $\frac{m}{n} \in [0, 1]$

From this definition it is possible to give general rules:

$$\dfrac{}{\vdash^{1}_{1} \Gamma, p, \bar{p}}\ (ax.)$$

$$\dfrac{}{\vdash^{0}_{1} \Delta}\ (\overline{ax.})$$

$$\frac{\vdash_n^m \Gamma, A, B}{\vdash_n^m \Gamma, A \vee B} \;(\vee)$$

$$\frac{\vdash_{n_1}^{m_1} \Gamma, A \qquad \vdash_{n_2}^{m_2} \Gamma, B}{\vdash_{n_1+n_2}^{m_1+m_2} \Gamma, A \wedge B} \;(\wedge)$$

This method will permit to keep track of the value at every stage of the proof.

**Example 1.2.** Let's take the same sequent considered before in example 1.1, but now using this decorated derivation.

$$\cfrac{\cfrac{\cfrac{\overline{\vdash_1^0 p, q}\;(\overline{ax.})}{\vdash_1^0 p \vee q}\;(\vee) \quad \cfrac{\overline{\vdash_1^1 p, \overline{p}}\;(ax.)}{\vdash_1^1 p \vee \overline{p}}\;(\vee)}{\vdash_2^1 (p \vee q) \wedge (p \vee \overline{p})}\;(\wedge) \quad \cfrac{\overline{\vdash_1^0 \overline{r}}\;(\overline{ax.}) \quad \overline{\vdash_1^0 \overline{t}}\;(\overline{ax.})}{\vdash_2^0 (\overline{r} \wedge \overline{t})}\;(\wedge)}{\vdash_4^1 (p \vee q) \wedge (p \vee \overline{p}) \wedge (\overline{r} \wedge \overline{t})}\;(\wedge)$$

The main difference between Example 1.1 and 1.2 is that in 1.2 is possible to read the fractional semantics value every time a new rule is introduced. For example the fractional semantics value of $\vdash (p \vee q) \wedge (p \vee \overline{p})$ will be $1/2 = 0.5$ as it's easily noticeable in the proof.

## 2 Framing beliefs into fractional semantics

Now that the general framework is presented, our aim is to see how to consider beliefs in the fractional semantics for classical logic. The idea is simple: a set of axioms, let's say $B$, that are considered true by an agent, is added to the system and every proposition in it is considered as true as tautologies. The philosophical idea behind this process is that an agent usually considers true their own beliefs.

Beliefs would be considered as deductively closed: this means that every deduction made using true beliefs will be considered true and this also means that the agent is a deductively ideal one. It is interesting to see what happens if the fractional semantics is provided with a system that is able to manage these new axioms, because it will be possible to obtain values greater than the ones that fractional semantics usually permits.

The idea for this kind of expansion was born thanks to the first of the three methods that Makinson used in [11] to "bridge the gap" between classical and non-monotonic logic, made by adding background assumptions. This kind of method was called *pivotal-assumption consequence* and permitted to infer more than classical logic permits thanks to a set of axioms that are added to the premises in every deduction.

Fractional semantics for classical logic updated with a set of new beliefs is different from pivotal-assumption consequence because of two main reasons. The first one is that Makinson used a classical two-valued semantics, whereas fractional semantics is a multi-valued interpretation. On the one hand *pivotal assumption consequence* would assign the value 0 if at least one of the axioms is neither a proper axiom nor a belief, on the other hand fractional semantics is provided by a system able to assign values greater than 0 when at least one of the top sequents is a tautology or a belief. The second reason is that, although both of them were born thanks to syntactical techniques, Makinson used an Hilbert-style approach, while fractional semantics uses the Gentzen-style one.

In order to add beliefs to the system, they must be atomic, if they are not, they must be decomposed, as it will be possible to see better.

**Definition 2.1** ($GS4_B$). Let $GS4$ as defined earlier, $GS4_B$ is defined as $GS4$ with a set of new axioms, namely $B = b_1, \ldots, b_n$, that represents a non-contradictory set of beliefs of an agent. Each $b_i$ ($1 \leq i \leq n$) must be atomic.

**Definition 2.2** ($\vdash_B$). If $\vdash$ is the closure relation of classical logic, $\vdash_B$ is defined as the closure relation of $GS4_B$.

Makinson [11] pointed out the problems arising when new axioms are added to the system. In fact, substitution is no longer acceptable when the system has the possibility to manage new non-logical axioms. The same result, even if it is not cited by Makinson, is due to the fact that classical logic is a Post-complete system and this means that, once a nontautological formula is added to the system, the new system will be inconsistent, unless structurality is dropped. The system loses the structurality because structurality and consistency are mutually excluding properties in classical logic with extra-logical axioms [13]. In the Makinson's formulation:

**Theorem 2.1.** *There is no supra-classical closure relation in the same language as classical $\vdash$ that is closed under substitution, except for $\vdash$ itself and the total relation i.e. the relation that relates every possible premises to every possible conclusion.*

*Proof.* See [11]. □

**Remark 2.1.** *From a conceptual viewpoint here is more fruitful to see why substitution is no longer acceptable into the system throughout an example. For instance let's consider $A = p \wedge q$. In order to add that we have to decompose it:*

$$\frac{\vdash_B p \quad \vdash_B q}{\vdash_B p \wedge q} \, (\wedge)$$

*Thus let's add the two clauses $p$ and $q$ to the set of beliefs. In classical logic it is possible to substitute $p \wedge q$ with a different clause, for example $r$ and obtain $\vdash_B r$, but $r$ is not one of the clauses added to the system and this means that $[\![r]\!]_B = 0$ while $[\![p]\!]_B = [\![q]\!]_B = 1$. It's even easier to understand it if the meaning of belief is analyzed. In everyday reasoning it is not possible to substitute beliefs with other beliefs at will and this is why substitution is not acceptable in the system.*

Now it's possible to go further into the formalization of the system: let's define the top sequent made from beliefs added to the system.

**Definition 2.3.** $top^b(\pi)$ : denotes the multiset of all and only $\pi$'s top-sequents introduced by a belief.

The new way to calculate the value of a sequent will be:

$$[\![A]\!]_B = \frac{top^b(\pi) + top^1(\pi)}{top^b(\pi) + top^1(\pi) + top^0(\pi)}$$

**Example 2.1.** Let's see an example taking the same sequent seen in example 1.1, but adding now the belief $\vdash_B (p \vee q) \wedge \overline{u}$. The first thing to do, in order to add the belief, is to decompose it and add that to the belief set.

$$\frac{\dfrac{\overline{\vdash_B p, q}\,(b_1)}{\vdash_B p \vee q}\,(\vee) \quad \overline{\vdash_B \overline{u}}\,(b_2)}{\vdash_B (p \vee q) \wedge \overline{u}}\,(\wedge)$$

From this it is possible to add the two beliefs $b_1 = p, q$ and $b_2 = \overline{u}$ to the belief set.

Let's take the same decomposition seen in example 1.1: it is possible to see where a belief is added to the system, namely $b_1$, because $\overline{u}$ doesn't appear in the sequent.

$$\cfrac{\cfrac{\cfrac{\vdash_B p, q}{\vdash_B p \vee q}(\vee)\ (b_1) \quad \cfrac{\cfrac{\vdash_B p, \overline{p}}{\vdash_B p \vee \overline{p}}(\vee)\ (ax.)}{}}{\vdash_B (p \vee q) \wedge (p \vee \overline{p})}(\wedge) \quad \cfrac{\cfrac{\vdash_B \overline{r}}{\vdash_B \overline{r} \wedge \overline{t}}(\overline{ax.}) \quad \cfrac{\vdash_B \overline{t}}{}(\overline{ax.})}{}(\wedge)}{\vdash_B (p \vee q) \wedge (p \vee \overline{p}) \wedge (\overline{r} \wedge \overline{t})}(\wedge)$$

This proof contains one identity axiom, one belief and two complementary axioms, so $[\![A]\!]_B$ is:

$$[\![A]\!]_B = \frac{top^b(\pi) + top^1(\pi)}{top^b(\pi) + top^1(\pi) + top^0(\pi)} = \frac{1+1}{1+1+2} = \frac{2}{4} = 0.5$$

Like in example 1.2, it is possible to see the same tree with multi valued system, adding a new rule:

$$\cfrac{}{\vdash^1_{1\ B} B}(\overline{b_i})$$

Where $B$ denote the set of belief, $B = b_1, \ldots, b_n$.

**Example 2.2.** Now it's possible to see Example 2.1 with the multi valued system.

$$\cfrac{\cfrac{\cfrac{\vdash^1_{1\ B} p, q}{\vdash^1_{1\ B} p \vee q}(\vee)\ (b_1) \quad \cfrac{\cfrac{\vdash^1_{1\ B} p, \overline{p}}{\vdash^1_{1\ B} p \vee \overline{p}}(\vee)\ (ax.)}{}}{\vdash^2_{2\ B} (p \vee q) \wedge (p \vee \overline{p})}(\wedge) \quad \cfrac{\cfrac{\vdash^0_{1\ B} \overline{r}}{\vdash^0_{2\ B} (\overline{r} \wedge \overline{t})}(\overline{ax.}) \quad \cfrac{\vdash^0_{1\ B} \overline{t}}{}(\overline{ax.})}{}(\wedge)}{\vdash^2_{4\ B} (p \vee q) \wedge (p \vee \overline{p}) \wedge (\overline{r} \wedge \overline{t})}(\wedge)$$

The only difference between Example 1.2 and 2.2 can be seen in the substitution with the value 1 instead of the 0 for the top axiom ⊢ $p, q$. In Figure 1 it is possible to see how the value of $A$ changed when was considered in the fractional semantics framework without beliefs and when one belief is added.

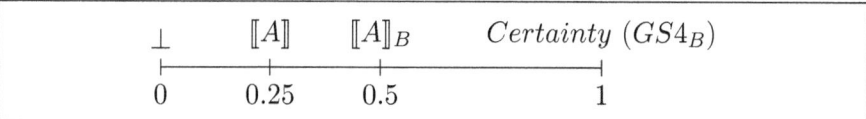

Figure 1: The values of $[\![A]\!]$ and $[\![A]\!]_B$.

It is worth noting that if this sequent was considered in classical logic, it would have different values changing by the value of the atomic formulas, but it would assume value 0 or 1. Something similar happens in Makinson pivotal assumption consequence, also if the belief set is the same that we have defined earlier, because a two valued logic is there considered.

## 2.1 Strong cut elimination

The last section pointed out that the agent is an ideal one and that they are aware of every deduction between beliefs. This means that the belief set is deductively closed: nothing that was not already in the set can be derived. In order to have a deductively closed belief set it is important that every combination of sentences, when it is possible, must be closed under cut and the new sentences obtained in this way will be added to the belief set.

In order to eliminate cut from $GS4_B$ the method is taken from [13], but it is simplified because of the nature of one-sided sequents. The method is the following:

1. turn each new belief $b_i$ added to the system in a conjunctive form: cnf($b_i$) = $b_1 \wedge \cdots \wedge b_n$ and add to the system each of the atomic formulas;

2. for each disjunctive formula, let's remove the application of ∨-rule: $b_1 \vee \cdots \vee b_n \equiv b_1, \ldots, b_n$;

3. let's remove copies of the same sequent;

4. let's remove identity sequents of the form ⊢ $\Gamma, p, \bar{p}$;

5. let's close under cut the belief set and add to the system the new formulas obtained in this way.

It is easy to show why the last step is so important. Suppose that an agent has a new belief: $A = (\bar{p} \wedge (\bar{t} \vee q)) \vee (t \wedge (\bar{t} \vee q))$. The first thing to do in order to add that belief is to transform $A$ in a conjunctive form: it is easy to show that it is equivalent to $\vdash (\bar{p} \vee t) \wedge (\bar{t} \vee q) \wedge (t \vee \bar{t} \vee q) \wedge (\bar{t} \vee q)$. Let's decompose it in a set of clauses: $\vdash \bar{p}, t, \vdash \bar{t}, q, \vdash t, \bar{t}, q, \vdash \bar{t}, q$ and remove one of the copies of $\vdash \bar{t}, q$ and the axiom $\vdash t, \bar{t}, q$. By the method presented earlier the agent has to add $(\bar{p} \vee t)$ and $(\bar{t} \vee q)$ to the system, but these beliefs are not cut free. To let them be cut free, it is necessary to close them under the cut.

$$\frac{\vdash \bar{p}, t \qquad \vdash \bar{t}, q}{\vdash \bar{p}, q} \; (cut)$$

From the last point of the method presented earlier, it is needed to add not only $\vdash \bar{p}, t$ and $\vdash \bar{t}, q$, but also $\vdash \bar{p}, q$. Let's see why: $[\![(\bar{p} \vee t) \wedge (\bar{t} \vee q)]\!]$ has value 1 if $B = \{(\bar{p}, t); (\bar{t}, q)\}$

$$\cfrac{\cfrac{\vdash^1_{1\,B} \bar{p}, t}{\vdash^1_{1\,B} \bar{p} \vee t} \; (\vee) \qquad \cfrac{\vdash^1_{1\,B} \bar{t}, q}{\vdash^1_{1\,B} \bar{t} \vee q} \; (\vee)}{\vdash^2_{2\,B} (\bar{p} \vee t) \wedge (\bar{t} \vee q)} \; (\wedge)$$

But what about $(\bar{p} \vee t) \wedge (\bar{t} \vee q) \wedge (\bar{p} \vee q)$?

**Remark 2.2.** $(\bar{p} \vee t) \wedge (\bar{t} \vee q) \wedge (\bar{p} \vee q)$ *is classically equivalent to* $(\bar{p} \vee t) \wedge (\bar{t} \vee q)$, *this means that they must have the same value because of stability, i.e.* $[\![(\bar{p} \vee t) \wedge (\bar{t} \vee q)]\!] = [\![(\bar{p} \vee t) \wedge (\bar{t} \vee q) \wedge (\bar{t} \vee q)]\!]$.

What happens if $B = \{(\bar{p}, t); (\bar{t}, q)\}$ is considered, instead of $B = \{(\bar{p}, t); (\bar{t}, q), (\bar{p}, q)\}$, obtained by adding also the belief closed under cut? Let's consider the decomposition of $(\bar{p} \vee t) \wedge (\bar{t} \vee q) \wedge (\bar{p} \vee q)$ within the set $B = \{(\bar{p}, t); (\bar{t}, q)\}$.

$$\cfrac{\cfrac{\vdash^1_{1\,B} \bar{p}, t}{\vdash^1_{1\,B} \bar{p} \vee t} (\vee) \quad \cfrac{\vdash^1_{1\,B} \bar{t}, q}{\vdash^1_{1\,B} \bar{t} \vee q} (\vee)}{\cfrac{\vdash^2_{2\,B} (\bar{p} \vee t) \wedge (\bar{t} \vee q)}{\vdash^2_{3\,B} (\bar{p} \vee t) \wedge (\bar{t} \vee q) \wedge (\bar{p} \vee q)}} (\wedge) \quad \cfrac{\vdash^0_{1\,B} \bar{p}, q}{\vdash^0_{1\,B} \bar{p} \vee q} (\vee) \; (\wedge)$$

This way a different value for the sequent is obtained and it must have the same value of $\vdash (\bar{p} \vee t) \wedge (\bar{t} \vee q)$.

This means that it is important to pay attention to frame not only the axioms obtained by the decomposition of the sequent, but also every formula closed under cut. In fact if the set $B = \{(\bar{p}, t); (\bar{t}, q), (\bar{p}, q)\}$ is considered, the original value is restored.

$$\cfrac{\cfrac{\vdash^1_{1B} \bar{p}, t}{\vdash^1_{1B} \bar{p} \vee t} (\vee) \quad \cfrac{\vdash^1_{1B} \bar{t}, q}{\vdash^1_{1B} \bar{t} \vee q} (\vee)}{\cfrac{\vdash^2_{2B} (\bar{p} \vee t) \wedge (\bar{t} \vee q)}{\vdash^3_{3B} (\bar{p} \vee t) \wedge (\bar{t} \vee q) \wedge (\bar{p} \vee q)} (\wedge)} \quad \cfrac{\cfrac{\vdash^1_{1B} \bar{p}, q}{\vdash^1_{1B} \bar{p} \vee q} (\vee)}{} (\wedge)$$

Thus the sequents have the same value:

$$[\![(\bar{p} \vee t) \wedge (\bar{t} \vee q)]\!]_B = [\![(\bar{p} \vee t) \wedge (\bar{t} \vee q) \wedge (\bar{p} \vee q)]\!]_B = 1$$

As it was showed, the cut is really important for a complete set of beliefs, but it is also necessary to see how the cut can be eliminated from the calculus.

**Elimination of cut** The elimination of cut in presence of proper axioms was firstly proposed by Girard [4], as noted by Avron [1], upgrading the Gentzen's standard cut elimination algorithm. The procedure here proposed, i.e., the decomposition of the formula, the add to the system and the cut of the formula to obtain all the derivations, owes a lot to the one presented in [13]. In the article, in fact, is proved that, for any cluster of extra-logical assumptions, there exists exactly one axiomatic extension of classical propositional logic that admits cut elimination. First of all it is possible to see that weakening it is admissible in $GS4_B$.

**Theorem 2.2** (Weakening admissibility in $GS4_B$). *For two atomic formulas $\vdash_B \Gamma$ and $\vdash_B \Delta$, $[\![\bigvee \Gamma \vee \bigvee \Delta]\!]_B \geq [\![\bigvee \Gamma]\!]_B$.*

*Proof.* To prove this is sufficient to consider a transformation of $\vdash_B$. In fact if $B = b_1, \ldots b_n$, then $\vdash_B \Gamma$ is equal to $\vdash \Gamma, \bar{b_1}, \ldots \bar{b_n}$ with the classical closure as pointed out in [11][1]. □

---

[1]In the text the two sided version of this transformation was used, so $\vdash_B \Gamma$ becomes $b_1, \ldots, b_n \vdash \Gamma$, but here because of the choice to use $GS4$ as main system, it is used the one-sided classically equivalent version $\vdash \Gamma, \bar{b_1}, \ldots, \bar{b_n}$.

It is possible to generalize this result for any context:

**Theorem 2.3.** *For any context $\Gamma$ and a formula $A$, such that $A$ is not contradictory with the set $B$, $[\![\bigvee \Gamma \vee A]\!]_B \geq [\![\Gamma]\!]_B$.*

*Proof.* By induction on the complexity of the formula $A$. □

The proof shows that this system is totally monotonic, also if it could seem counterintuitive for a set of belief., because maybe if an agent has a belief $p$ it is strange to believe also $p \wedge q$, but this is due to the fact that this work is based on a classical framework and so the fractional value of $[\![p]\!]_B$ assumes the same of $[\![p \wedge q]\!]_B$.

**Theorem 2.4** (Strong cut elimination of $GS4_B$). *The cut rule is redundant when added to $GS4_B$.*

*Proof.* Similar to the one proposed in [13]. □

The set of beliefs can be "completed" through cut or without that. This means that $GS4_B$ is a cut-free system, because it is an axiomatic extension of classical logic. By the way, the use of cut can alter the fractional semantics value as shown in [12]. Thanks to theorem 2.4 the algorithm presented in section 2.1 can be transformed in an algorithm without the presence of cut. As a corollary of the strong cut elimination it can be obtained:

**Theorem 2.5** (Uniqueness of axiomatization in $GS4_B$). *For any cluster of axioms in the set of beliefs $B$ the axiomatization is unique.*

*Proof.* See [13]. □

# 3 Conclusions

The present research may be considered as an attempt to create a bridge between logic and the real world, nonetheless retaining consistency and decidability.

The main difference between classical approach and the fractional semantics one is either philosophical and technical. By a technical point of view is possible to distinguish between different levels of contradiction. Fractional semantics is able to distinguish between what's inside truthfulness and falsity, without losing the rigorous approach to proof theory, proving the cut elimination and the others fundamental properties.

The system $GS4_B$ is able to consider beliefs and logical axioms together. The fractional system, in fact, does not have the symmetry between tautologies and contradictions and it is helpful to talk about something that is uncertain such as beliefs.

**Further researches** The current implementation we have can be further extended to explore other possibilities. For instance, we could investigate the impact of limiting the set of beliefs by applying fractional semantics to non-monotonic logics or non-classical logics in general. This could lead to a deeper understanding of how these logics work and what their limitations might be, which could in turn inform the development of new logics or the improvement of existing ones. Additionally, exploring the application of fractional semantics to other areas of computer science, such as artificial intelligence or natural language processing, could yield promising results and open up new avenues for research. Overall, there are many exciting directions in which this implementation could be expanded, and we look forward to exploring them in future work.

In a future work, the Lottery Paradox will be resolved while maintaining closure under conjunction between beliefs that will show a very interesting application of beliefs and fractional semantics.

# References

[1] Avron, A. Gentzen-Type Systems, Resolution and Tableaux. *Journal Of Automated Reasoning.* **10** pp. 265-281 (1993)

[2] Foley, R. The Epistemology of Belief and the Epistemology of Degrees of Belief. *American Philosophical Quarterly.* **29**, 111-124 (1992)

[3] Gärdenfors, P. & Makinson, D. Revisions of Knowledge Systems Using Epistemic Entrenchment. *Proceedings Of The 2nd Conference On Theoretical Aspects Of Reasoning About Knowledge.* pp. 83-95 (1988)

[4] Girard, J. Proof and types. (Cambridge university press,1989)

[5] Grove, A. Two Modellings for Theory Change. *Journal Of Philosophical Logic.* **17**, 157-170 (1988)

[6] Halpern, J. Reasoning about Uncertainty. (MIT Press,2003)

[7] Hawthorne, J. The Lockean Thesis and the Logic of Belief. *Degrees Of Belief.* pp. 49-74 (2009)

[8] Hughes, D. A minimal classical sequent calculus free of structural rules. (arXiv,2005)

[9] Kleene, S. Mathematical logic. (John Wiley and Sons,1967)

[10] Leitgeb, H. The Stability of Belief: How Rational Belief Coheres with Probability. (Oxford University Press,2017)

[11] Makinson, D. Bridges from classical to nonmonotonic logic. (Lightning Source, Milton Keynes,2005)

[12] Piazza, M. & Pulcini, G. Fractional Semantics for Classical Logic. *Review Of Symbolic Logic.* **13**, 810-828 (2020)

[13] Piazza, M. & Pulcini, G. Uniqueness of axiomatic extensions of cut-free classical propositional logic. *Logic Journal Of The IGPL.* **24**, 708-718 (2016,10)

[14] Piazza, M., Pulcini, G. & Tesi, M. Fractional-valued modal logic. *The Review Of Symbolic Logic.* pp. 1-22 (2021,8)

[15] Smullyan, R. First-order logic. (Courier corporation,1995)

[16] Troelstra, A. & Schwichtenberg, H. Basic Proof Theory. (Cambridge University Press,2000)

# Ignorance and Its Formal Limits

Ekaterina Kubyshkina
*Logic, Uncertainty, Computation and Information Group (LUCI)*
*Research Center for the Philosophy of Technology (PhilTech)*
*Department of Philosophy, University of Milan*
`ekaterina.kubyshkina@unimi.it`

Mattia Petrolo
*Centre for Philosophy of Science of the University of Lisbon*
*(CFCUL)*
*Federal University of ABC*
`mpetrolo@fc.ul.pt`

**Abstract.** Over the last decades, several works in epistemic logic focused on the challenge of modeling ignorance. Given that this notion is considered as intricate and multi-layered, one of the difficulties in this endeavor lies in the lack of consensus regarding the specific form of ignorance being targeted for modeling. The aim of this article is to provide an overview of diverse frameworks for representing ignorance, delineate the particular forms of ignorance captured by each framework, and propose formal definitions for certain types of ignorance that lack explicit treatment in the existing literature.

**Keywords:** ignorance representation, epistemic logic, three-valued modal logic.

## 1 Introduction

Several current works in epistemic logic focus on finding a way to model the notion of ignorance (see, e.g., [18], [25], [5], [11], [19], [1]). One of the difficulties in achieving this task is that there is no agreement on exactly which notion one is trying to model. The Standard View on ignorance defines it simply as non-knowledge[1]. From this perspective, ignorance conceived as

---
[1] See, e.g., [21] and the references therein.

non-knowledge can be modeled using standard epistemic logic, as suggested in [17]. In particular, given the epistemic modality $K$ for knowledge representation, an agent is 'ignorant that' $\phi$ if the agent does not know that $\phi$, i.e., $\neg K\phi$ holds (see [12]). However, such a representation trivializes the notion of ignorance. In particular, if one accepts that knowledge is factive[2], as it is usually the case in epistemic logic, it is easy to see that an agent is ignorant of any false proposition.

1. $\neg \phi$;
2. $\neg K\phi$ (by $T$ and contraposition).

However, it sounds counterintuitive to claim that an agent is ignorant that "Milan is not in Italy" simply because this proposition is false.

Moreover, it seems that expressing ignorance simply as not-knowing does not permit one to distinguish among various forms of ignorance. Unlike knowledge, ignorance manifests itself in many distinct forms. For instance, one can be ignorant of $\phi$ because they disbelieve $\phi$ (i.e., they believe that $\phi$ is false); or one can be ignorant of $\phi$ because they suspend their belief on whether $\phi$ is true (so-called 'suspending ignorance'); or one can be ignorant of $\phi$ because they neither believe $\phi$, nor disbelieve $\phi$, nor suspend their belief on $\phi$ (so-called 'deep ignorance')[3]. Modelling ignorance via $\neg K$ does not permit one to formally capture the important differences between these situations. From this perspective, it seems promising to provide a formal framework for each type of ignorance. The aim of this article is to give an overview of various frameworks for ignorance representation, indicate which type of ignorance is captured by each framework, and propose formal definitions of some types of ignorance that are currently missing in the literature.

## 2 Formal Representations

In this section, we survey three general approaches based on different theoretical assumptions about ignorance and discuss their limits in representing this notion in general. The first approach aims to represent *ignorance whether*, as explicitly proposed in [18]. The second approach aims to capture *ignorance of unknown truths* (see [25]). The third approach, introduced in

---

[2]That is, the system validates the axiom $T$: $K\phi \to \phi$

[3]These situations do not provide an exhaustive list of forms of ignorance. This is an interesting task investigated by various epistemologists (see, e.g., [24]), but it is not the aim of this article.

[19], represents *disbelieving ignorance*, although the authors initially refer to this type of ignorance as 'factive ignorance.' The common feature of these approaches is the representation of ignorance via a primitive modality. An advantage of this strategy lies in providing a formal setting for considering specific types of ignorance without necessarily reducing it to the standard epistemic framework, thereby capturing the minimal conditions for representing ignorance.

## 2.1 Ignorance whether

The first definition we consider is that of van der Hoek & Lomuscio [18], who take ignorance to be 'not knowing whether' (that is, ignorance whether) and provide a system for representing it. Semantically, they use standard Kripke semantics, in which ignorance is represented via a primitive modality (henceforth denoted as $I^w$), definable via $K$ as $\neg K\phi \wedge \neg K\neg\phi$. It should be noted that this definition coincides with the contingency operator $\nabla$, indicating that the authors consider ignorance as an epistemic counterpart of contingency.

In what follows, we utilize standard Kripke semantics, without modification to standard presentations. A *frame* is a pair $\langle W, R \rangle$ where $W$ is a non-empty set of possible worlds, and $R \subseteq W \times W$ is an accessibility relation. A *model* is a pair $\langle \mathcal{F}, v \rangle$ where $\mathcal{F}$ is frame and $v : var \to \mathcal{P}(W)$ is a valuation function. The primitive modality $I^w$ is thus defined in a model $\mathcal{M}$ in a world $w$ as follows:[4]

**Definition 1** (Operator $I^w$). *Let* $\mathcal{M} = \langle W, R, v \rangle$.

- $\mathcal{M}, w \models I^w \phi$ *iff there exists $w'$ such that $Rww'$ and $\mathcal{M}, w' \models \phi$ and there exists $w''$ such that $Rww''$ and $\mathcal{M}, w \models \neg\phi$.*

This type of ignorance representation is further developed in [5]. The neighbourhood approach is provided in [6], [7], and [9]. An approach via labelled sequent calculus is provided in [16]. All the systems introduced are proved to be sound and complete.

Several observations should be made concerning the use of $I^w$ as an operator representing ignorance. First, it does not represent ignorance *in general*, but a particular case of ignorance, *ignorance whether*. For instance, to represent the so-called *second-order ignorance*, that is, one is ignorant

---

[4]The definitions of a formula being true in a model in a world are standard for Boolean cases and thus are omitted. The definitions of truth and validity in a model and in a frame are also standard, so they are also omitted.

whether one is ignorant whether $\phi$ (see [12]), a double application of $I^w$ is needed: $I^w I^w \phi$. Secondly, it does not permit distinguishing between two distinct types of ignorance: deep and suspending ignorance. In particular, as mentioned in the Introduction, suspending ignorance is the type of ignorance in which an agent is aware of the content of the proposition but suspends judgment on the truth-value of the proposition. Deep ignorance means that the agent is unaware of the content of the proposition. The semantic definition of $I^w \phi$ presupposes that there is an accessible world in which $\phi$ holds and an accessible world in which $\neg\phi$ holds. How should this be interpreted from the agent's point of view? An intuitive answer to this is that an agent considers $\phi$ as possibly true and possibly false, thus suspending judgement on $\phi$. However, this shows that $I^w$ does not represent deep ignorance. Another possible interpretation of the definition of $I^w \phi$ in natural language is that the agent considers neither $\phi$ nor $\neg\phi$ as true propositions. This applies to both suspending and deep ignorance. However, in this case, the semantics proposed for $I^w$ is too unexpressive to distinguish between deep and suspending ignorance.

## 2.2 Ignorance of unknown truths

The second definition of ignorance operator is introduced by Steinsvold [25, 26]. In this case, ignorance of $\phi$ is considered as if $\phi$ is true, but unknown to the agent. This kind of ignorance is represented semantically using a primitive modality denoted by $I^u$, which can be defined using $K$ as $\phi \wedge \neg K \phi$. This modality coincides with the non-consistency operator $\bullet$, indicating that Steinsvold considers ignorance as an epistemic counterpart of non-consistency. On Kripke models, the operator is defined as follows:

**Definition 2** (Operator $I^u$). Let $\mathcal{M} = \langle W, R, v \rangle$.

- $\mathcal{M}, w \models I^u \phi$ iff $\mathcal{M}, w \models \phi$ and there exists $w'$ such that $Rww'$ and $\mathcal{M}, w' \models \neg \phi$.

This type of ignorance representation is further developed in [22] and [14].[5] The neighbourhood approach is provided in [15]. An approach via labelled sequent calculus is provided in [16]. All the systems introduced are proved to be sound and complete.

---

[5]The formal analysis provided in [22] and [14] does not explicitly concern ignorance representation. However, taking into account that $I^u$ is an epistemic counterpart of the non-consistency operator, the results for the logics containing (non-)consistency operator also hold for ignorance of unknown truths.

Similarly to the case of ignorance whether, it should be noted that $I^u$ does not represent ignorance in general. For instance, the following principle is valid in any Kripke model.

**Proposition 1.** *For any $\mathcal{M} = \langle W, R, v \rangle$ and any $w \in W$:*

- $\mathcal{M} \models I^u \phi \to I^u I^u \phi$.

*Proof.* Let $\mathcal{M} = \langle W, R, v \rangle$ and there exists $w \in W$ s.t. $\mathcal{M}, w \not\models I^u \phi \to I^u I^u \phi$. Then, (i) $\mathcal{M}, w \models I^u \phi$ and (ii) $\mathcal{M}, w \not\models I^u I^u \phi$. From (i) we have that there exists $w'$ s.t. $Rww'$ and $\mathcal{M}, w' \models \neg \phi$, which means that (iii) $\mathcal{M}, w' \models \neg I^u \phi$. From (i) and (ii) we have that (iv) for all $w''$ such that $Rww''$, $\mathcal{M}, w'' \models I^u \phi$, which contradicts (iii). □

This proposition states that if an agent is ignorant of a truth, they are always ignorant of their own ignorance. This shows that the $I^u$ modality cannot represent cases of ignorance of which the agent is not ignorant (for instance, suspending ignorance).

Additionally, similar to the case of $I^w$, one can question the natural language interpretation of the semantic clause for $I^u$. This reveals that either deep ignorance is not representable using the $I^u$ operator or that it is not possible to distinguish deep ignorance from other forms of ignorance using this modality.

## 2.3 Disbelieving ignorance

A third option for ignorance representation is provided by Kubyshkina & Petrolo [19]. For reasons that will become clear later, we call this type of ignorance *disbelieving ignorance*, and denote $I^d$ the primitive modality associated with it. Unlike $I^w$ and $I^u$, $I^d$ is not reducible to the standard $K$ modality, as it is shown in [13]. The definition of $I^d$ on standard Kripke models follows.

**Definition 3** (Operator $I^d$). Let $\mathcal{M} = \langle W, R, v \rangle$.

- $\mathcal{M}, w \models I^d \phi$ iff for all $w' \neq w$ if $Rww'$ then $\mathcal{M}, w' \models \neg \phi$ and $\mathcal{M}, w \models \phi$.

This type of ignorance representation is further developed and equipped with neighbourhood semantics in [13]. An approach via labelled sequent calculus is provided in [16]. All the systems introduced are proved to be sound and complete.

The semantic clause for $I^d\phi$ to be true in a world $w$ states that $\phi$ is true in $w$, but considered false in all worlds accessible from $w$ (which are not $w$ itself). This is why we call this type of ignorance disbelieving ignorance: for everything that the agent considers to be true, $\phi$ is false.

It is clear that the case of disbelieving ignorance only represents one form of ignorance. Similar to the $I^u$ operator, the $I^d$ operator cannot represent cases of conscious ignorance, as suggested by the following proposition:

**Proposition 2.** *For any* $\mathcal{M} = \langle W, R, v\rangle$ *and any* $w \in W$:

- $\mathcal{M} \models I^d \to I^d I^d \phi$.

Furthermore, as was the case for $I^w$ and $I^u$, $I^d$ does not permit one to represent deep ignorance, or, at the very least, it does not enable the distinction between deep ignorance and other types of ignorance.

We have shown that all the three settings for representing ignorance are unable to represent at least one form of ignorance. Therefore, we express some skepticism regarding the possibility of representing ignorance in general through a single modality. However, we do not view this as a drawback. Epistemologists refine their analysis of the notion of ignorance by considering its various types (see, e.g., [24]). Thus, we believe that various forms of ignorance should be represented in distinct ways, allowing us to capture formally the difference between the corresponding types of ignorance. Specifically, the three operators, $I^w$, $I^u$, and $I^d$, represent three types of ignorance, none of which constitute an exclusive case of deep ignorance. Deep ignorance is characterized by an absence of an attitude towards both proposition and its negation. From this perspective, it seems that a crucial aspect of the formal representation of deep ignorance is the characerization of this absence. In classical propositional logic, which we used as the underlying setting for Kripke semantics, the presence of a proposition corresponds to its truth in a world. Unfortunately, this does not allow us to distinguish whether the 'absence of a proposition' means that an agent considers its falsity or does not consider it at all. The truth conditions for all the three operators, $I^w$, $I^u$, and $I^d$, include a case in which a proposition $\phi$, of which the agent is ignorant, does not hold in some accessible world (or in all of them, in the case of $I^d$). Due to the use of classical propositional logic, this leads to the truth of $\neg\phi$ in all these worlds. However, stating the truth of $\neg\phi$ is equivalent to stating the falsity of $\phi$. Therefore, the use of classical logic does not allow us to distinguish the 'absence of $\phi$' from its falsity, as these two notions collapse. In what follows, we explore a possible setting for eliminating this ambiguity between falsity and absence of a truth value by introducing a three-valued setting for ignorance representation. Other options can also be considered for this purpose, such as exploiting awareness functions (see [4]). This proposal is a promising alternative, but, for the sake of the current article, we leave this option for further investigation.

## 3 Three-valued setting

The first attempts to represent ignorance in a many-valued setting can be found in [3] and [20]. In [3], a three-valued setting is proposed to represent what is called 'severe' ignorance, which refers to ignorance of a proposition resulting from an unfamiliarity with its meaning. To achieve this, the authors introduce three-valued Kripke models, replacing classical propositional logic with three-valued Bochvar's logic (see [2]). By interpreting the third additional value, $\frac{1}{2}$, as 'meaningless,' the authors define the truth of severe ignorance as follows: an agent is severely ignorant of $\phi$ in a world $w$ if $\phi$ is meaningless in all worlds accessible from $w$. Since $\frac{1}{2}$ is interpreted as meaningless, this setting does not distinguish between falsity and absence, and thus is not suitable for representing deep ignorance.

In [20], a logic for *excusable ignorance* is proposed. Following [23], the authors consider that fully excusable ignorance is constituted of two cases: deep ignorance and disbelieving ignorance. Semantically, the authors define excusable ignorance as a variation of the $I^d$ operator interpreted on Kripke models with possibly incomplete worlds.[6]

**Definition 4** (Models with possibly incomplete worlds). *A model $\mathcal{M} = \langle \mathcal{F}, v \rangle$ is called a model with possibly incomplete worlds whenever $\mathcal{F}$ is a Kripke frame and $v$ is a valuation function such that, for each atomic proposition $p$, $v(p) \to \mathcal{P}(W)$, $v(\neg p) \to \mathcal{P}(W)$, and $v(p) \cap v(\neg p) = \emptyset$.*

**Definition 5** (Excusable ignorance). *Let $\mathcal{M} = \langle \mathcal{F}, v \rangle$ be a model with possibly incomplete worlds.*

- $\mathcal{M}, w \models I^e \phi$ *iff for all $w' \neq w$ if $Rww'$ then $\mathcal{M}, w' \not\models \phi$ and $\mathcal{M}, w \models \phi$.*
- $\mathcal{M}, w \models \neg I^e \phi$ *otherwise.*

---

[6]We do not consider definitions of non-modal connectives here, which can be introduced according to a chosen three-valued logic.

In accordance with this semantics, there are three mutually exclusive and exhaustive possibilities for each propositional variable $p$: either it is true in a world $w$ ($\mathcal{M}, w \models p$), or it is false in a world $w$ ($\mathcal{M}, w \models \neg p$), or it is neither true, nor false ($\mathcal{M}, w \not\models p$ and $\mathcal{M}, w \not\models \neg p$). However, the authors have chosen to define excusable ignorance in a bivalent way: either $I^e \phi$ is true in a world, or it is false. This is due to the underlying assumption that either an agent is excusably ignorant, or they are not, that is, an agent cannot be neither excusable nor not-excusable for their actions. The truth conditions for excusable ignorance are thus as follows: an agent is excusably ignorant of $\phi$ in $w$ iff for all worlds accessible from $w$ that are not $w$ itself, $\phi$ is not true (and thus it is either false or neither true nor false in these worlds), and $\phi$ is true in $w$. Such a formulation makes it explicit that the $I^e$ operator includes the case of deep ignorance – whenever $\phi$ is neither false nor true in all worlds accessible from $w$ – and the case of disbelieving ignorance – whenever $\phi$ is false in all these worlds. Keeping this inclusion in mind, we can now propose a definition of deep ignorance derived from the definition of $I^e$.

**Definition 6** (Deep ignorance). *Let $\mathcal{M} = \langle \mathcal{F}, v \rangle$ be a model with possibly incomplete worlds.*

- $\mathcal{M}, w \models I^{deep} \phi$ *iff for all $w' \neq w$ if $Rww'$ then $\mathcal{M}, w' \not\models \phi$ and $\mathcal{M}, w' \not\models \neg\phi$, and $\mathcal{M}, w \models \phi$.*

- $\mathcal{M}, w \models \neg I^{deep} \phi$ *otherwise.*

We can also redefine other types of ignorance by using Kripke models with possibly incomplete worlds. For instance, the following definition eliminates any ambiguity in defining disbelieving ignorance: in order to be disbelievingly ignorant of $\phi$, the agent considers $\phi$ to be a false proposition based on all that she knows.

**Definition 7** (Disbelieving ignorance). *Let $\mathcal{M} = \langle \mathcal{F}, v \rangle$ be a model with possibly incomplete worlds.*

- $\mathcal{M}, w \models I^d \phi$ *iff for all $w' \neq w$ if $Rww'$ then $\mathcal{M}, w' \models \neg\phi$ and $\mathcal{M}, w \models \phi$.*

It is also useful to reconsider the $I^w$ and $I^u$ operators in a three-valued setting. In particular, by adjusting $I^w$, one can distinguish a particular type of ignorance – the suspending one – in which an agent considers a proposition but is unsure about its truth-value.

**Definition 8** (Suspending ignorance). *Let $\mathcal{M} = \langle \mathcal{F}, v \rangle$ be a model with possibly incomplete worlds.*

- $\mathcal{M}, w \models I^s \phi$ iff there exists $w'$ such that $Rww'$ and $\mathcal{M}, w' \models \phi$ and there exists $w''$ such that $Rww''$ and $\mathcal{M}, w \models \neg \phi$.

One can also adjust the $I^u$ operator to represent ignorance resulting from incomplete information. There are, of course, different options for defining this case of ignorance.

**Definition 9** (Ignorance from incomplete information - 1). *Let $\mathcal{M} = \langle \mathcal{F}, v \rangle$ be a model with possibly incomplete worlds.*

- $\mathcal{M}, w \models I^{incl} \phi$ iff $\mathcal{M}, w \models \phi$ and there exists $w'$ such that $Rww'$ and $\mathcal{M}, w' \not\models \phi$.

In accordance with this definition, an agent is ignorant of a truth iff there is at least one possibility from the agent's perspective that the proposition is not true. However, there might be at least two reasons for this kind of situation. First, the agent might explicitly consider that the proposition is false (that is, there exists a world validating the negation of this proposition). Second, the agent might have incomplete access to information about the proposition (that is, there is at least one world in which neither the proposition nor its negation is true). This second option seems to correspond better to situations of ignorance caused by incomplete information. For this reason, we propose the following alternative to Def. 9.

**Definition 10** (Ignorance from incomplete information - 2). *Let $\mathcal{M} = \langle \mathcal{F}, v \rangle$ be a model with possibly incomplete worlds.*

- $\mathcal{M}, w \models I^{inc2} \phi$ iff $\mathcal{M}, w \models \phi$ and there exists $w'$ such that $Rww'$, $\mathcal{M}, w' \not\models \phi$, and $\mathcal{M}, w' \not\models \neg \phi$.

# 4 Conclusion

We have considered various approaches to representing ignorance and found that none of them can fully capture ignorance in general, but only specific types of ignorance. We argue that this does not limit ignorance representation but rather provides a more nuanced understanding of this complex notion, similar to the analysis carried out in epistemology. We then focused on the challenge of representing deep ignorance in a formal setting and proposed a framework that employs Kripke semantics with possibly incomplete worlds. This three-valued approach enabled us to rethink other types of ignorance and clarify their formal definitions. A promising development of our study is the introduction of a unified system with multiple modalities for ignorance as primitive operators, allowing for a uniform analysis of the interactions between different types of ignorance. The first steps in this direction can be found in [8], [10], and [16].

# Acknowledgements

We would like to thank two anonymous reviewers for their comments. The research of Ekaterina Kubyshkina is funded under the "Foundations of Fair and Trustworthy AI" Project of the University of Milan. Ekaterina Kubyshkina is further funded by the Department of Philosophy "Piero Martinetti" of the University of Milan under the Project "Departments of Excellence 2023-2027" awarded by the Ministry of University and Research (MUR). Mattia Petrolo acknowledges the financial support of the FCT – Fundação para a Ciência e a Tecnologia (2022.08338.CEECIND; R&D Unit Grants UIDB/00678/2020 and UIDP/00678/2020) and the French National Research Agency (ANR) through the Project ANR-20-CE27-0004.

# References

[1] Aldini, A., Graziani, P., Tagliaferri, M. (2023). "A hierarchical characterization of ignorance in epistemic logic [Special Issue]." *Argumenta*, 8(2): 373-386.

[2] Bochvar, D. A. (1938). "On a three-valued logical calculus and its application to the analysis of the paradoxes of the classical extended functional calculus." *Recueil Mathématique [Matematicheskiy Sbornik]*, 4(46): 2, 287-308.

[3] Bonzio, S., Fano, V., Graziani, P. (2023). "A logical modeling of severe ignorance." *Journal of Philosophical Logic*, online first: https://doi.org/10.1007/s10992-022-09697-x.

[4] Fagin, R., Halpern J. Y. (1988). "Belief, Awareness and Limited Reasoning." *Artificial Intelligence*, 34(1): 39-76.

[5] Fan, J., Wang, Y., van Ditmarsch, H. (2015). Contingency and knowing whether. *The Review of Symbolic Logic*, 8(1): 75-107.

[6] Fan, J., can Ditmarsch, H. (2015). "Neighborhood contingency logic." In M. Banerjee and S. Krishna (eds), *Logic and Its Application*, vol. 8923 of Lecture Notes in Computer Science, pages 88-99.

[7] Fan, J. (2018). "Neighborhood contingency logic: A new perspective." *Studies in Logic*, 11(4): 37-55.

[8] Fan, J. (2019). "Bimodal logics with contingency and accident." *Journal of Philosophical Logic*, 48: 425-445.

[9] Fan, J. (2019). "A family of neighborhood contingency logics." *Notre Dame Journal of Formal Logic*, 60(4): 683-699.

[10] Fan, J. (2021). "Bimodal logic with contingency and accident: Bisimulation and axiomatizations." *Logica Universalis*, 15: 123-147.

[11] Fano, V., Graziani, P. (2020). A working hypothesis for the logic of radical ignorance. *Synthese*, 199(1-2): 601-616.

[12] Fine, K. (2018). Ignorance of ignorance. *Synthese*, 195(9): 4031-4045.

[13] Gilbert, D., Kubyshkina, E., Petrolo, M., Venturi, G. (2021). "Logics of ignorance and being wrong." *Logic journal of the IGPL*, Online First, pp. 1-16 DOI: 10.1093/jigpal/jzab025.

[14] Gilbert, D., Venturi, G. (2016). "Reflexive insensitive modal logics." *The Review of Symbolic Logic*, 9: 167-180.

[15] Gilbert, D., Venturi, G. (2017). "Neighborhood semantics for logics of unknown truths and false beliefs." *The Australasiona Journal of Logic*, 14(1): 246-267.

[16] Girlando, M., Kubyshkina, E., Petrolo, M. (2022). "A proof-theoretic approach to ignorance." Booklet of abstract of the *LATD 2022 and Mosaic Kick off Meeting*, http://logica.dipmat.unisa.it/LATD+MOSAIC/abstracts/MOSAIC2022_paper_8472.pdf.

[17] Hintikka, J. (1962). *Knowledge and Belief*. Ithaca, Cornell University Press.

[18] van der Hoek, W., Lomuscio, A. (2004). A logic for ignorance. *Electronic Notes in Theoretical Computer Science*, 85(2): 117-133.

[19] Kubyshkina, E., Petrolo, M. (2021). A logic for factive ignorance. *Synthese*, 198: 5917-5928.

[20] Kubyshkina, E., Petrolo, M., Pereira, M. K. (2022). "Ignorance as an excuse, formally." Presentation at Logic, Uncertainty, Computation, Information Group's seminar, University of Milan, https://www.youtube.com/watch?v=8Ihdxfr5qls.

[21] Le Morvan, P., Peels, R. (2016). The nature of ignorance: two views. *The epistemic dimensions of ignorance*, Peels, R., Blaauw, M. (eds), Cambridge University Press, pages 12-32.

[22] Marcos, J. (2005). "Logic of essence and accident." *Bulletin of the Section of Logic*, 34(1): 43-56.

[23] Peels, R. (2014). "What kind of ignorance excuses? Two neglected issues." *Philosophical Quarterly*, 64(256): 478-496.

[24] Peels, R. (2019). "Asserting ignorance." In *The Oxford Handbook of Assertion*, S. Goldberg (ed.), pp. 604-624.

[25] Steinsvold, C. (2008a). A note on logics of ignorance and borders. *Notre Dame Journal of Formal Logic*, 49(4): 385-392.

[26] Steinsvold, C. (2008b). Completeness for various logics of essence and accident. *Bulletin of the Section of Logic*, 37(2): 93-101.

# A note on schematic validity and completeness in Prawitz's semantics

Antonio Piccolomini d'Aragona
*Institute of Philosophy, Czech Academy of Sciences*
piccolomini@flu.cas.cz

**Abstract.** I discuss two approaches to monotonic proof-theoretic semantics. In the first one, which I call SVA, consequence is understood in terms of existence of valid arguments. The latter involve the notions of argument structure and justification for arbitrary non-introduction rules. In the second approach, which I call Base Semantics, structures and justifications are left aside, and consequence is defined outright over background atomic theories. Many (in)completeness results have been proved relative to Base Semantics, the question being whether these can be extended to SVA. By limiting myself to a framework with classical meta-logic, I prove correctness of classical logic on Base Semantics, and show that this result adapts to SVA when justifications are allowed to be choice-functions over atomic theories or unrestricted reduction systems of argument structures. I also point out that, however, if justifications are required to be more schematic, correctness of classical logic over SVA may fail, even with classical logic in the meta-language. This seems to reveal that the way justifications are understood may be a distinguishing feature of different accounts of proof-theoretic validity.
**Keywords:** proof-theoretic semantics, completeness, schematicity

## 1 Introduction

Prawitz's semantics, an instance of proof-theoretic semantics [1, 13], has come in basically two forms: *semantics of valid arguments* – SVA, see e.g.

---

I thank Cesare Cozzo, Ansten Klev, Ivo Pezlar, Thomas Piecha, Paolo Pistone, Dag Prawitz, Peter Schroeder-Heister and Will Stafford for helpful comments. Work on this article was supported by grant LQ300092101 from the Czech Academy of Sciences

[9] – and *theory of grounds* – ToG, see e.g. [10]. Here, I shall focus mostly on SVA, and I will only occasionally refer to ToG.

SVA is based on the notion of *valid argument*. The latter is inspired by Prawitz's normalisation results for Gentzen's Natural Deduction [8], stating that derivations for $\Gamma \vdash A$ can be transformed, through suitable reductions, to derivations for $\Gamma^* \subseteq \Gamma \vdash A$ without *detours*. A detour is given by a formula which occurs both as conclusion of an introduction, and as a major premise of an elimination.

In intuitionistic logic, Prawitz's theorems imply what Schroeder-Heister called the *fundamental corollary* [12]: $A$ is a theorem iff there is a closed derivation of $A$ ending by an introduction. This may confirm Gentzen's claim that introductions fix meaning, while eliminations are unique functions of the introductions [2].

Prawitz's well-known *inversion principle* [8] that a by-introduction proof of $A$ contains (part of) what is needed for drawing consequences of $A$, can thus undergo a semantic reading: arbitrary inferences may be *justified* by transformations on proofs of the major premises. An argument in general can be said to be valid when its arbitrary inferences are so justified. This might happen relative to some underlying atomic proof-system, meant to determine non-logical meanings, or hold irrespective of such systems, and thus qualify as logical.

In the SVA-inspired framework of some other authors, which following Sandqvist [11] we may call *Base Semantics*, the role of justifications and proof-structures is limited to atomic proof-systems. Validity of arguments is replaced by a consequence relation among formulas, which may once again be relative to given atomic proof-systems, or hold on all such systems, i.e. be logical.

Works in the Base Semantics tradition have shown that intuitionistic completeness depends on peculiar features of atomic proof-systems – see [5] for an overview, while later examples are [7, 14, 15, 16]. Building upon a result of [7], I here prove incompleteness of intuitionistic logic with respect to a variant of SVA, and so show that, under a certain reading, Base Semantics copes with an approach where justifications and proof-structures are not disregarded. However, I also point out that the passage from Base Semantics to SVA is non-trivial, since the adaptation of the result of [7] to SVA seems to force an understanding of reductions as *non-schematic* functions from and to proof-structures.

In particular, I shall argue that a more "schematic" understanding of reductions blocks certain proofs of soundness of classical logic, which instead obtain when reductions are non-schematic. This may mean that a substantive part of the constructive nature of Prawitz's semantics stems from a certain understanding of reductions, rather than just from the fact that the approach is proof-based.

Although SVA and Base Semantics can be (and have been) developed for full first-order logic, intuitionistic (in)completeness is mostly discussed at a propositional level. I accordingly limit myself to propositional logic. Moreover, in proving intuitionistic incompleteness, I use classical logic at the meta-level. This is however sufficient for raising my point.

The article is structured as follows. In Section 2 I give an outline of (a variant of) SVA. In Section 3 I define (a variant of) Base Semantics, and prove incompleteness of intuitionistic logic with respect to it – specifically, soundness of excluded middle with respect to it. In Section 4 I adapt the incompleteness proof to SVA in such a way as to stick to Prawitz's original idea of a semantics where justifications and proof-structures do play an active role. In Section 5 I discuss the incompleteness result of Section 4 with reference to a notion of justification understood on different degrees of strength. In the concluding remarks, I finally suggest some tentative proposals for accounting for a "schematic" notion of reduction, in terms of restrictions on the form of open valid arguments, of Pezlar selector, and of linearity conditions on replacement of variable parts in proof-structures.

## 2 SVA (over a base)

Prawitz's normalisation obtains through reduction functions from and to Gentzen's Natural Deduction derivations, whose iteration eventually normalise application arguments while keeping the conclusion and not expanding the assumptions-set. For disjunction detours, such a function is e.g.

$$
\cfrac{\cfrac{\mathscr{D}_1}{A_i}\ (\vee_I), i=1,2}{A_1 \vee A_2} \quad \begin{matrix}[A_1] & [A_2]\\ \mathscr{D}_2 & \mathscr{D}_3 \\ B & B\end{matrix} \ (\vee_E) \qquad \stackrel{\phi_\vee}{\Longrightarrow} \qquad \begin{matrix}\mathscr{D}_1\\ [A_i]\\ \mathscr{D}_{i+1}\\ B\end{matrix}
$$

This is brought to a semantic level by SVA, through a suitable generalisation of the notions of derivation and reduction towards, respectively, the notions of argument structure and justification, relative to a background language $\mathscr{L}$ which will be here given as follows.

**Definition 1.** *The grammar of the* language $\mathscr{L}$ *is*

$$X := p \mid \bot \mid X \wedge X \mid X \vee X \mid X \to X.$$

The $p$-s and $\bot$ are atoms, $\bot$ being a constant symbol for the absurd. As done above, I use capital Latin letter as variables for formulas of $\mathscr{L}$, while capital Greek letters will be used as variables for sets of formulas. Negation is defined by

$$\neg A \stackrel{def}{=} A \to \bot.$$

Just like in Model Theory, also in SVA we have structures which fix the meaning of the non-logical components of $\mathscr{L}$, and thus serve as induction-base for the definition of the semantic core-notions. Unlike Model Theory, however, these structures are not sets (or values) onto which non-logical components are mapped, but atomic systems, i.e. sets of rules that establish the semantic behaviour of atoms – and thus determine constructively the meaning of the components they involve. Atomic systems can be defined in different ways, depending on different desiderata one may have – see [6] – and yielding different outcomes for validity-related concepts such as completeness – see [5]. Here, I will understand them as sets of *production rules*, i.e. I will not allow premises to be lower-level rules, nor will I allow for discharge of assumptions at the atomic level.

**Definition 2.** *An* atomic base $\mathfrak{B}$ *over $\mathscr{L}$ is a (countable) set of rules*

$$\frac{A_1, ..., A_n}{B}$$

*where $n \geq 0$, $A_i, B$ are atoms of $\mathscr{L}$ and $A_i \neq \bot$ ($i \leq n$).*

Derivations on $\mathfrak{B}$ are defined in a usual inductive way. The derivability relation on $\mathfrak{B}$ is written $\vdash_\mathfrak{B}$, while the derivations-set of $\mathfrak{B}$ is written DER$_\mathfrak{B}$. I always require bases to be consistent.

**Definition 3.** $\mathfrak{B}$ *is* consistent *iff $\nvdash_\mathfrak{B} \bot$.*

**Definition 4.** *An* argument structure *over $\mathscr{L}$ is a pair $\langle T, f \rangle$ such that $T$ is a tree whose nodes are either empty, in which case they are always top-nodes, or formulas of $\mathscr{L}$, and $f$ is a function defined on a subset $\Gamma$ of non-empty top-nodes of $T$ such that, for every $A \in \Gamma$, $f(A)$ is below $A$ in $T$.*

Given $\mathscr{D} = \langle T, f \rangle$ where $T$ has top-nodes $\Gamma$ and root $A$, I call $\Gamma$ the *assumptions* of $\mathscr{D}$ and $A$ the *conclusion* of $\mathscr{D}$ – thus $f$ is an assumptions-discharge.

**Definition 5.** $\mathscr{D}$ is closed *iff all its assumptions are discharged, otherwise it is* open.

Where $\Gamma$ is the set of the open assumptions of a $\mathscr{D}$ with conclusion $A$, I shall say that $\mathscr{D}$ is an argument structure *from $\Gamma$ to $A$*, and I shall indicate it by

$$\begin{array}{c} \Gamma \\ \mathscr{D} \\ A \end{array}$$

**Definition 6.** *Given $\mathscr{D}$ from $\Gamma$ to $A$ and $\sigma$ a function such that, for every $B \in \Gamma$, $\sigma(B)$ is a (closed) argument structure with conclusion $B$, a (closed) $\sigma$-instance $\mathscr{D}^\sigma$ of $\mathscr{D}$ is the argument structure obtained from $\mathscr{D}$ by replacing every $B \in \Gamma$ with $\sigma(B)$.*

**Definition 7.** *An* inference *is a triple $\langle \langle \mathscr{D}_1, ..., \mathscr{D}_n \rangle, A, \delta \rangle$, where $\delta$ is an indication of assumptions which may be discharged by the inference. The argument structure associated to the inference, indicated by*

$$\frac{\mathscr{D}_1, ..., \mathscr{D}_n}{A} \delta$$

*is obtained by conjoining the trees of the $\mathscr{D}_i$-s through a root-node $A$, and by expanding the assumptions-discharges of the $\mathscr{D}_i$-s according to $\delta$. A* rule *is a set of inferences, whose elements are called* instances of the rule.

I shall assume that rules can be described schematically, e.g. standard introduction rules in Gentzen's Natural Deduction,

$$\frac{A \quad B}{A \wedge B} (\wedge_I) \qquad \frac{A_i}{A_1 \vee A_2} (\vee_I), i = 1, 2 \qquad \frac{[A]}{A \to B} (\to_I)$$

**Definition 8.** $\mathscr{D}$ is canonical *iff it is associated to an instance of an introduction rule, otherwise it is* non-canonical.

**Definition 9.** *Given a rule $R$, a* justification *of $R$ is a constructive function $\phi$ defined on the set of the argument structures $\mathbb{D}$ associated to some sub-set of $R$ such that, for every $\mathscr{D} \in \mathbb{D}$,*

- *$\mathscr{D}$ is from $\Gamma$ to $A \Longrightarrow \phi(\mathscr{D})$ is from $\Gamma^* \subseteq \Gamma$ to $A$, and*
- *for every $\sigma$, $\phi$ is defined on $\mathscr{D}^\sigma$ and $\phi(\mathscr{D}^\sigma) = \phi(\mathscr{D})^\sigma$.*

For example, the reduction $\phi_\vee$ for eliminating maximal disjunctions from Natural Deduction derivations, which has been defined at the beginning of this section, can be understood semantically as a justification showing that elimination of disjunction ($\vee_E$) can be safely removed, *salva* provability, when its major premise has been obtained by ($\vee_I$). Observe that $\phi_\vee$ needs not be defined on all the applications of ($\vee_E$). Via Definition 12 below, applications whose major premise is proved via ($\vee_I$) only suffice for showing the rule to be justified in the sense hinted at above.

I take the notion of (*immediate*) *sub-structure* of $\mathscr{D}$ to be clear enough at this point, so I will not define it explicitly. The same I will do with the notion of *substitution* of a sub-structure $\mathscr{D}^*$ by an argument structure $\mathscr{D}^{**}$ in an argument structure $\mathscr{D}$, written $\mathscr{D}[\mathscr{D}^{**}/\mathscr{D}^*]$ – observe that this already occurs in Definition 6. The application of substitution may require re-indexing the discharge functions associated to the argument structures, but I shall not deal with these details here.

**Definition 10.** *Given a set of justifications* $\mathfrak{J}$, $\mathscr{D}$ *immediately reduces to* $\mathscr{D}^*$ *relative to* $\mathfrak{J}$, *written* $\mathscr{D} \leq_i^\mathfrak{J} \mathscr{D}^*$, *iff* $\mathscr{D} = \mathscr{D}^*$ *or, for some sub-structure* $\mathscr{D}^{**}$ *of* $\mathscr{D}$ *and some* $\phi \in \mathfrak{J}$, $\phi$ *is defined on* $\mathscr{D}^{**}$ *and* $\mathscr{D}^* = \mathscr{D}[\phi(\mathscr{D}^{**})/\mathscr{D}^{**}]$. $\mathscr{D}$ *reduces to* $\mathscr{D}^*$ *relative to* $\mathfrak{J}$, *written* $\mathscr{D} \leq^\mathfrak{J} \mathscr{D}^*$, *iff there is a sequence* $\mathscr{D} = \mathscr{D}_1 \leq_i^\mathfrak{J} \mathscr{D}_2 \leq_i^\mathfrak{J} ... \leq_i^\mathfrak{J} \mathscr{D}_{n-1} \leq_i^\mathfrak{J} \mathscr{D}_n = \mathscr{D}^*$.

**Definition 11.** *An argument is a pair* $\langle \mathscr{D}, \mathfrak{J} \rangle$.

**Definition 12.** $\langle \mathscr{B}, \mathfrak{J} \rangle$ *is valid on* $\mathfrak{V}$ *iff*

- $\mathscr{D}$ *is closed* $\Longrightarrow$
  - *the conclusion of* $\mathscr{D}$ *is an atom* $\Longrightarrow$ $\mathscr{D} \leq^\mathfrak{J} \mathscr{D}^*$ *for* $\mathscr{D}^* = \langle T, \emptyset \rangle$ *and* $T \in DER_\mathscr{B}$ *closed;*
  - *the conclusion of* $\mathscr{D}$ *is not an atom* $\Longrightarrow$ $\mathscr{D} \leq^\mathfrak{J} \mathscr{D}^*$ *for* $\mathscr{D}^*$ *closed canonical with immediate sub-structures valid on* $\mathfrak{B}$ *when paired with* $\mathfrak{J}$;

- $\mathscr{D}$ *is open from* $\Gamma$ *to* $A$ $\Longrightarrow$ *for every* $\sigma$, *every* $B \in \Gamma$ *and every extension* $\mathfrak{J}^+$ *of* $\mathfrak{J}$, *if* $\langle \sigma(B), \mathfrak{J}^+ \rangle$ *is valid on* $\mathfrak{B}$, *then* $\langle \mathscr{D}^\sigma, \mathfrak{J}^+ \rangle$ *is valid on* $\mathfrak{B}$.

Let us prove that $\langle \mathscr{D}, \{\phi_\vee\} \cup \mathfrak{J}_2 \cup \mathfrak{J}_3 \rangle$ is valid on any $\mathfrak{B}$, where $\mathscr{D}$ is the structure

$$\cfrac{A_1 \vee A_2 \quad \begin{matrix} [A_1] \\ \mathscr{D}_2 \\ B \end{matrix} \quad \begin{matrix} [A_2] \\ \mathscr{D}_3 \\ B \end{matrix}}{B} (\vee_E)$$

and $\langle \mathscr{D}_2, \mathfrak{J}_2 \rangle$ and $\langle \mathscr{D}_3, \mathfrak{J}_3 \rangle$ are open valid on $\mathfrak{B}$ from $A_1$ and $A_2$ respectively to $B$. By the second clause of Definition 12, we must prove that, for every $\langle \mathscr{D}_1, (\{\phi_\vee\} \cup \mathfrak{J}_2 \cup \mathfrak{J}_3)^+ \rangle$ valid on $\mathfrak{B}$ where $\mathscr{D}_1$ is closed and has conclusion $A_1 \vee A_2$, we have that $\langle \mathscr{D}[\mathscr{D}_1/A_1 \vee A_2], (\{\phi_\vee\} \cup \mathfrak{J}_2 \cup \mathfrak{J}_3)^+ \rangle$ is valid on $\mathfrak{B}$. Since $\langle \mathscr{D}_1, (\{\phi_\vee\} \cup \mathfrak{J}_2 \cup \mathfrak{J}_3)^+ \rangle$ is closed valid on $\mathfrak{B}$, by the first clause of Definition 12, there is a closed canonical $\mathscr{D}_1^*$ with conclusion $A_1 \vee A_2$ such that $\mathscr{D}_1 \leq_{(\{\phi_\vee\} \cup \mathfrak{J}_2 \cup \mathfrak{J}_3)^+} \mathscr{D}_1^*$, and $\langle \mathscr{D}_1^*, (\{\phi_\vee\} \cup \mathfrak{J}_2 \cup \mathfrak{J}_3)^+ \rangle$ is valid on $\mathfrak{B}$. Also, $\mathscr{D}_1 \leq_{(\{\phi_\vee\} \cup \mathfrak{J}_2 \cup \mathfrak{J}_3)^+} \mathscr{D}_1^*$ obviously implies $\mathscr{D} \leq_{(\{\phi_\vee\} \cup \mathfrak{J}_2 \cup \mathfrak{J}_3)^+} \mathscr{D}^*$, where $\mathscr{D}^*$ is closed of the form

$$\cfrac{\cfrac{\mathscr{D}_1^*}{\cfrac{A_i}{A_1 \vee A_2}(\vee_I),\ i=1,2} \quad \cfrac{[A_1]}{\mathscr{D}_2}{B} \quad \cfrac{[A_2]}{\mathscr{D}_3}{B}}{B}(\vee_E)$$

We must show that $\langle \mathscr{D}^*, (\{\phi_\vee\} \cup \mathfrak{J}_2 \cup \mathfrak{J}_3)^+ \rangle$ is valid on $\mathfrak{B}$, since this implies $\langle \mathscr{D}[\mathscr{D}_1/A_1 \vee A_2], (\{\phi_\vee\} \cup \mathfrak{J}_2 \cup \mathfrak{J}_3)^+ \rangle$ valid on $\mathfrak{B}$ which in turn, by the arbitrary choice of $\langle \mathscr{D}_1, (\{\phi_\vee\} \cup \mathfrak{J}_2 \cup \mathfrak{J}_3)^+ \rangle$, implies our result. Since $\phi_\vee \in (\{\phi_\vee\} \cup \mathfrak{J}_2 \cup \mathfrak{J}_3)^+$, we have that $\mathscr{D}^* \leq_{(\{\phi_\vee\} \cup \mathfrak{J}_2 \cup \mathfrak{J}_3)^+} \mathscr{D}^{**}$, with $\mathscr{D}^{**}$ closed of the form

$$\mathscr{D}_1^*$$
$$[A_i]$$
$$\mathscr{D}_{i+1}$$
$$B$$

Since we assumed $\langle \mathscr{D}_{i+1}, \mathfrak{J}_{i+1} \rangle$ to be open valid on $\mathfrak{B}$, by the second clause of Definition 12 $\langle \mathscr{D}^{**}, (\{\phi_\vee\} \cup \mathfrak{J}_2 \cup \mathfrak{J}_3)^+ \rangle$ is closed valid on $\mathfrak{B}$, whence we are done.

## 3 Base semantics

Since argumental validity essentially depends on atomic provability on the base, on reducibility of non-canonical configurations to suitable introduction forms, and on assumptions-closure in the open-arguments case, we may decide to prune Definition 12 by just focusing on formulas, i.e. by dropping argument structures and justifications out.

This leads to Base Semantics, which I give in a somewhat simplified version as what [7] calls *non-extension* semantics.

**Definition 13.** $\Gamma \models_\mathfrak{B} A$ *iff:*

(a) $\Gamma = \emptyset \Longrightarrow$

   (At) $A$ is an atom $\Longrightarrow \vdash_\mathfrak{B} A$;

   ($\wedge$) $A = B \wedge C \Longrightarrow \models_\mathfrak{B} B$ and $\models_\mathfrak{B} C$;

   ($\vee$) $A = B \vee C \Longrightarrow \models_\mathfrak{B} B$ or $\models_\mathfrak{B} C$;

   ($\to$) $A = B \to C \Longrightarrow B \models_\mathfrak{B} C$;

(b) $\Gamma \neq \emptyset \Longrightarrow (\models_\mathfrak{B} \Gamma \Longrightarrow \models_\mathfrak{B} A)$, where $\models_\mathfrak{B} \Gamma$ means $\models_\mathfrak{B} B$ for every $B \in \Gamma$.

**Definition 14.** $\Gamma \models A$ iff, for every $\mathfrak{B}$, $\Gamma \models_\mathfrak{B} A$.

We can now give an easy incompleteness proof for intuitionistic logic – IL. In fact, we can even prove soundness of classical logic (which is expected by the use of the latter in the meta-language).

**Proposition 1.** *For every $\mathfrak{B}$, $\models_\mathfrak{B} A \vee \neg A$.*

*Proof.* For $\mathfrak{B}$ arbitrary, by using classical logic in the meta-language, either $\models_\mathfrak{B} A$ or $\not\models_\mathfrak{B} A$. If $\models_\mathfrak{B} A$, then $\models_\mathfrak{B} A \vee \neg A$ by ($\vee$). If instead $\not\models_\mathfrak{B} A$, then $A \models_\mathfrak{B} \bot$ holds vacuously by (b). Hence, $\models_\mathfrak{B} \neg A$ by ($\to$), and $\models_\mathfrak{B} A \vee \neg A$ by ($\vee$). The result now follows from the arbitrariness of $\mathfrak{B}$. $\square$

**Proposition 2.** $\models A \vee \neg A$.

**Theorem 1.** *There are $\Gamma$ and $A$ such that $\Gamma \models A$ and $\Gamma \not\vdash_{IL} A$.*

Except for the reference to atomic bases – which is in a sense peculiar to Prawitzian approaches – this is essentially the same as what one obtains in BHK-semantics with classical meta-language.

The crucial results of intuitionistic incompleteness to be found in literature on Base Semantics are of course much more significant than the one provided by Theorem 1 – and, accordingly, their proofs require a much more fine-grained framework than the one I put forward here. In [7], incompleteness of intuitionistic logic is e.g. referred to *Harrop rule*, whose validity via Definition 14 is proved without using classical logic in the meta-language.[1]

---

[1]It should be however remarked that, in [7], the rule from premise $A \to (B \vee C)$ to conclusion $(A \to B) \vee (A \to C)$ – with no restrictions on $A$ – is proved to be valid according to Definition 14, by using classical meta-logic and (b) in Definition 13. This can be easily adapted to the SVA-framework along the lines of Section 4 below. I shall not dwell upon this here, as the use of classical meta-logic validates excluded middle, which immediately implies intuitionistic incompleteness (both in Base Semantics and in a suitably modified version of SVA).

The adaptation of the Base Semantics framework to SVA that I present in Section 4 below can be extended to these more refined approaches; likewise, the critical remarks about the principles of Base Semantics I provide in Section 5 below also holds for more detailed accounts – both in Base Semantics and in adaptations of it to SVA. Thus, the toy-example that, in the (trivialising) context of a classical meta-language, I discuss in Sections 4 and 5, are enough for raising my points. A comprehensive treatment of the issue can be developed in future works.[2]

## 4 A kind of intuitionistic SVA-incompleteness

The obvious strategy – also suggested by [7] – for moving from Base Semantics to SVA, is that of starting from the following equivalence:

(EQ) $\Gamma \models_\mathfrak{B} A$ iff there is $\langle \mathscr{D}, \mathfrak{J} \rangle$ from $\Gamma$ to $A$ valid on $\mathfrak{B}$.

Condition (a) in Definition 13 holds in SVA under (EQ), so we may just translate Proposition 1 via (EQ), after observing that, with classical logic in the meta-language, condition (b) in Definition 13 also holds under (EQ) in SVA – proof omitted.

However, I prefer to stick to a strict SVA-formulation, which will allow me to remark some points that might be concealed when argument structures and justifications are entirely dropped out.

**Proposition 3.** *For every $\mathfrak{B}$, there is a closed argument for $A \vee \neg A$ valid on $\mathfrak{B}$.*

*Proof.* Let $\mathfrak{B}$ be arbitrary. By reasoning classically, either there is a closed valid argument for $A$, or there is not. Suppose there is not, and consider

$$\mathscr{D}^* = \frac{A}{\bot}$$

---

[2]Let me just remark that the fundamental results proved in [7] require some additional principles, called *Import* and *Export*, and possibly higher-level rules at the atomic level – i.e. atomic rules where the discharge of assumptions is allowed. The *Import* principle says that, roughly, some assumptions may generate new atomic rules: $\Gamma \models_\mathfrak{B} A \Leftrightarrow \models_{\mathfrak{B} \cup \Sigma^\Gamma} A$. Together with higher-level atomic rules, this yields validity of Harrop rule via Definition 14 in non-extensions semantics. The *Export* principle says instead that, roughly, (first-level) atomic rules generate new assumptions: $\Gamma \models_{\mathfrak{B} \cup \Sigma} A \Leftrightarrow \Gamma, \Delta^\Sigma \models_\mathfrak{B} A$. This yields validity of Harrop rule via Definition 14 for *extension* semantics, namely Base Semantics where condition (b) in Definition 13 is given a monotonic form: $\Gamma \models_\mathfrak{B} A \Leftrightarrow$ for all $\mathfrak{C} \supseteq \mathfrak{B}, (\models_\mathfrak{C} \Gamma \Rightarrow \models_\mathfrak{C} A)$.

Then, $\langle \mathscr{D}^*, \emptyset \rangle$ is vacuously valid on $\mathfrak{B}$, and so is $\langle \mathscr{D}^{**}, \emptyset \rangle$, where $\mathscr{D}^{**}$ is

$$\dfrac{\dfrac{[A]^1}{\bot}}{A \to \bot} (\to_I), 1$$

Then, $\langle \mathscr{D}, \{\kappa_1\} \rangle$ is valid on $\mathfrak{B}$, where $\mathscr{D}$ is

$$\overline{A \vee \neg A}$$

and $\kappa_1$ is the constant function defined by

$$\mathscr{D} \quad \Longrightarrow \quad \dfrac{\overset{\mathscr{D}^{**}}{\neg A}}{A \vee \neg A}(\vee_I)$$

Suppose there is a closed $\langle \mathscr{D}^{***}, \mathfrak{J} \rangle$ for $A$ valid on $\mathfrak{B}$. Then $\langle \mathscr{D}, \{\kappa_2\} \cup \mathfrak{J} \rangle$ is valid on $\mathfrak{B}$, where $\kappa_2$ is the constant function defined by

$$\mathscr{D} \quad \Longrightarrow \quad \dfrac{\overset{\mathscr{D}^{***}}{A}}{A \vee \neg A}(\vee_I)$$

The result now follows by arbitrariness of $\mathfrak{B}$. □

A similar result for classical *reduction ad absurdum* can be found in [9]. Observe that, in Proposition 3, the role of (b) in Definition 13 is played by empty sets of justifications or by constant functions. Definition 14 via (EQ) now gives a kind incompleteness of IL.

**Definition 15.** $\Gamma \models_\Delta A$ *iff, for every $\mathfrak{B}$, there is $\langle \mathscr{D}, \mathfrak{J} \rangle$ from $\Gamma$ to $A$ valid on $\mathfrak{B}$.*

**Proposition 4.** $\models_\Delta A \vee \neg A$.

**Theorem 2.** *There are $\Gamma$ and $A$ such that $\Gamma \models_\Delta A$ and $\Gamma \nvdash_{IL} A$.*

## 5 Schematicity

When read against the original formulation of SVA in [9], the proof of Proposition 3 implies a number of aspects which may be so modified as to obtain a different reading of $\models_\Delta$ than the one I have proposed in the previous section. First, we may invert the quantifiers in Definition 15.

**Definition 16.** $\Gamma \models_\Delta^* A$ *iff, for some* $\langle \mathscr{D}, \mathfrak{J} \rangle$ *from* $\Gamma$ *to* $A$ *and every* $\mathfrak{B}$, $\langle \mathscr{D}, \mathfrak{J} \rangle$ *is valid on* $\mathfrak{B}$.

Our proof for Theorem 2 does not directly apply to $\models_\Delta^*$. Yet, the proof of Proposition 3 shows that, for every $\mathfrak{B}$, there is a set of justifications $\mathfrak{J}^\mathfrak{B}$ such that $\langle \mathscr{D}, \mathfrak{J}^\mathfrak{B} \rangle$ valid on $\mathfrak{B}$, where $\mathscr{D}$ is the axiom $A \vee \neg A$, i.e.

$$\mathfrak{J}^\mathfrak{B} = \begin{cases} \{\kappa_1\} & \text{if } \not\models_\mathfrak{B} A \\ \{\kappa_2\} \cup \mathfrak{J} & \text{otherwise} \end{cases}$$

with $\mathfrak{J}$ as required in the proof of Proposition 3. One idea for applying Theorem 2 to $\models_\Delta^*$ may thus be that of modifying Definition 9 so as to have justifications defined, not from argument structures to argument structures, but from subsets of $\mathbb{S} \times \mathbb{B}$ to arguments, where $\mathbb{S}$ is the sets of the argument structures over $\mathscr{L}$ and $\mathbb{B}$ is the class of atomic bases. In this way, we may set

$$\mathtt{Ch}\langle \mathscr{D}, \mathfrak{B} \rangle = \langle \mathscr{D}, \mathfrak{J}^\mathfrak{B} \rangle.$$

Ch would thus behave like a choice function picking the right justifications set on each base, so $\langle \mathscr{D}, \mathtt{Ch} \rangle$ is valid on every $\mathfrak{B}$ – we remark that this strategy requires a number of changes in the formal apparatus of Section 2, which eventually lead to ToG where, roughly, justifications of rules are embedded into inference steps – see [4].

Another alternative would be to put all the $\mathfrak{J}^\mathfrak{B}$-s into the justifications set for $\mathscr{D}$, so that $\langle \mathscr{D}, \bigcup_{\mathfrak{B} \in \mathbb{B}} \mathfrak{J}^\mathfrak{B} \rangle$ be valid on every $\mathfrak{B}$. Contrarily to the previous solution, this move does not require we touch Definition 9 at all.

A final option, which is similar to that where we take the union-set of the justifications set of each base, but which, similarly to the choice function strategy, involves a modification of Definition 9, amounts to following [12], namely, replacing, so to say, the justifications with their graph.[3] Let me give in this case more details than I have done so far. First of all, the modified Definition 9 now reads as follows.

---

[3]There are here some subtleties which I cannot deal with. In the approach inspired by [12], one normally allows for alternative justifications for one and the same argument structure – a possibility I have left open in Definition 9. In general,

**Definition 17.** A reduction *is a pair* $\langle \mathscr{D}^1, \mathscr{D}^2 \rangle$ *where* $\mathscr{D}^2$ *has the same conclusion, and at most the same assumptions as* $\mathscr{D}^1$. *A r-system is a set of reductions. A r-sequence* $\langle \mathscr{D}_1^1, \mathscr{D}_1^2 \rangle, ..., \langle \mathscr{D}_n^1, \mathscr{D}_n^2 \rangle$ *such that, for every* $i \leq n$, $\mathscr{D}_{i-1}^2 = \mathscr{D}_i^1$, *is said to be from* $\mathscr{D}_1^1$ *to* $\mathscr{D}_n^2$.

Let us modify accordingly Definitions 10, 11 and 12.

**Definition 18.** *Given a r-system* $\Sigma$, $\mathscr{D}$ *reduces to* $\mathscr{D}^*$ *relative to* $\Sigma$, *written* $\mathscr{D} \leq^\Sigma \mathscr{D}^*$, *iff* $\Sigma$ *contains a r-sequence from* $\mathscr{D}$ *to* $\mathscr{D}^*$.

**Definition 19.** *An* argument *is a pair* $\langle \mathscr{D}, \Sigma \rangle$.

**Definition 20.** $\langle \mathscr{D}, \Sigma \rangle$ *is SH-valid on* $\mathfrak{B}$ *iff the same conditions as in Definition 12 hold, with* $\mathfrak{J}$ *and* $\leq^\mathfrak{J}$ *replaced by* $\Sigma$ *and* $\leq^\Sigma$ *respectively.*

It is now easy to see that the following obtains.

**Proposition 5.** *For every* $\mathfrak{B}$, *there is closed* $\langle \mathscr{D}, \Sigma \rangle$ *for* $A \vee \neg A$ *SH-valid on* $\mathfrak{B}$.

*Proof.* Basically the same as proof of Proposition 3, where the relevant reductions are $\langle \mathscr{D}, \kappa_1(\mathscr{D}) \rangle$ and $\langle \mathscr{D}, \kappa_2(\mathscr{D}) \rangle$, respectively. □

This shows that, for every $\mathfrak{B}$, there is a r-system $\Sigma_\mathfrak{B}$ such that $\langle \mathscr{D}, \Sigma_\mathfrak{B} \rangle$ is valid on $\mathfrak{B}$ – where $\mathscr{D}$ is the axiom $A \vee \neg A$. Let us now give the following definition of logical validity.

**Definition 21.** $\Gamma \models_\Delta^{SH} A$ *iff, for some* $\langle \mathscr{D}, \Sigma \rangle$ *from* $\Gamma$ *to* $A$ *and every* $\mathfrak{B}$, $\langle \mathscr{D}, \Sigma \rangle$ *is SH-valid on* $\mathfrak{B}$.

Thus we have what follows.

**Proposition 6.** $\models_\Delta^{SH} A \vee \neg A$.

*Proof.* Take $\langle \mathscr{D}, \Sigma \rangle$ with $\mathscr{D}$ the axiom $A \vee \neg A$, and $\Sigma = \bigcup_{\mathfrak{B} \in \mathbb{B}} \Sigma_\mathfrak{B}$. □

**Theorem 3.** *There are* $\Gamma$ *and* $A$ *such that* $\Gamma \models_\Delta^{SH} A$ *and* $\Gamma \not\vdash_{IL} A$.

---

one can show that from any justifications set one can extract what, in Definition 17 below, I call a r-sequence; if alternative justifications are permitted, the inverse holds too. Something similar obtains concerning the relationship between validity of Definition 12 and what I call SH-validity in Definition 20 below: if alternative justifications are not permitted, then the implication holds only for closed argument structures, otherwise one has an equivalence between the notions.

There is however a clear sense in which neither **Ch**, nor $\bigcup_{\mathfrak{B}\in\mathbb{B}}\mathfrak{J}_{\mathfrak{B}}$ or $\bigcup_{\mathfrak{B}\in\mathbb{B}}\Sigma_{\mathfrak{B}}$ can be said to be or to consist of *schematic* justifications, where by schematic we may provisionally mean that the justification can be expressed as a *rule for rewriting argument structures*, similar to the reduction for removing disjunction detours that we have seen at the beginning of Section 2. **Ch** is obviously not so, since it is not even defined on argument structures only, but on pairs of arguments structures *and* bases. Even if it is in a sense schematic, it hence differs significantly in spirit from $\phi_\vee$.

On the other hand, $\bigcup_{\mathfrak{B}\in\mathbb{B}}\mathfrak{J}^{\mathfrak{B}}$ contains of course a schematic justification, i.e. $\kappa_1$. But the $\kappa_2$ we take on each base cannot be assumed to be schematic. This is a constant function which points the axiom $A \vee \neg A$ to a fixed closed argument structure for $A \vee \neg A$ obtained by introducing disjunction below a closed argument structure for $A$, where the latter is given. Clearly, we have no guarantee that the closed argument structure thereby obtained has a form that is invariant over all bases, and hence that $\kappa_2$ can be described as a rewriting scheme. Also, we have no information at all about $\mathfrak{J}$, i.e. the justifications sets for the closed argument structure for $A$ (if any). For essentially the same reasons, we may not be able to describe $\Sigma$ in proof of Proposition 6.

Under a suitable description of this notion of schematicity, we may thus modify $\models^*_\Delta$ in Definition 16, so as to obtain a further notion of logical consequence.

**Definition 22.** $\Gamma \models^s_\Delta A$ *iff, there is $\mathscr{D}$ from $\Gamma$ to $A$ such that, for some schematic $\mathfrak{J}$ and every $\mathfrak{B}$, $\langle \mathscr{D}, \mathfrak{J} \rangle$ is valid on $\mathfrak{B}$.*

This somewhat stricter reading may cope with constructivist desiderata. The validity of excluded middle may be no longer provable, even with classical meta-logic.

# 6 Conclusion

Prawitz's original aim was that of developing a constructive semantics which accounted for IL [9]. Investigations into proof-theoretic semantics have instead shown that IL is in general *not* complete, or that it is so only on condition that atomic bases undergo certain constraints – see [5] for an overview, while more recent findings are [7, 14, 15, 16]. However, these incompleteness (or restricted completeness) results are normally referred to a Base Semantics approach, where justifications and proof-structures are dropped out. In some cases, such results can be adapted to the SVA approach, but this may require understanding reductions in a non-schematic way, say, as choice-functions defined also on bases, or as non-recursive sets of justifications or of reductions sequences in Schroeder-Heister's sense. If we require more schematicity, the adaptation may fail, e.g., even with classical logic in the meta-language, we may no longer be able to prove the validity of excluded middle in the object-language. This would speak in favour of Prawitz's original semantic project being constructive in two strictly interrelated ways, namely, not just because SVA is based on a notion of valid argument (i.e. of proof), but *also* because this notion is given in terms of reductions which amount to concrete rewriting rules for proof-structures.

Of course, the question now becomes how schematicity should be more precisely defined. I will conclude by hinting at two possible strategies for dealing with this topic.

First, one may consider the possibility of relaxing the notion of open validity, i.e. an open valid argument $\langle \mathscr{D}, \mathfrak{J} \rangle$ is such that $\mathscr{D}$ reduces to canonical form relative to $\mathfrak{J}$ even when its assumptions *are not* replaced by closed structures which be valid relative to expansions of $\mathfrak{J}$. Observe that this *does not hold* in SVA. For example, take the base given by the rules

$$\frac{}{p} \qquad \frac{p}{q}$$

and consider the rule

$$\frac{r}{q \vee s}$$

justified by a $\phi$ such that

$$\frac{\dfrac{\overline{p}}{r}}{q \vee s} \quad \Longrightarrow \quad \frac{\dfrac{\overline{p}}{q}}{q \vee s} \,(\vee_I)$$

Then, $\langle \mathscr{D}, \phi \rangle$ is valid on our base, where $\mathscr{D}$ is

$$\frac{\dfrac{\overline{p}}{r}}{q \vee s}$$

It may nonetheless be the case that open structures whose assumptions have some *peculiar feature* may be computable to canonical form relative to some base-independent justifications, namely, justifications which be schematic in the broad sense understood here. A solution of this kind may come from Pezlar's selector [3] for the *Split rule*.

$$\frac{A \to B \vee C}{(A \to B) \vee (A \to C)}$$

with $A$ Harrop formula. Pezlar's selector is based on the idea that open proofs of the implicational premise involve sufficient computational content for them to reduce to canonical form under trivial *ad hoc* proofs of the antecedent.

Alternatively, one may put a sort of linearity constraint on justifications, like the one put by Definition 9 on applications of justifications to instances of open structures. We may require the same for the structures which the justification is defined on. If we go back to the reduction for disjunction detours, we see that it consists of a *constant* part, i.e. the order of the formulas on the left- and right-hand side of the arrow, and of a *variable* part, given by variables for argument structures. Let us express this as a linear operation $\vee E$ on typed proof-objects as happens in ToG, and let us indicate with inj a function for obtaining a proof of $A_1 \vee A_2$ from one of either $A_1$ or $A_2$. Then, the rule enjoys the following linearity condition: for any $f_1^1, f_2^1 : (A_1)B$, $f_1^2, f_2^2 : (A_2)B$ and $x_1^i, x_2^i : A_i$ ($i = 1, 2$),

$$\vee E(\text{inj}(x_1^i)[x_2^i/x_1^i], f_1^1(y_1)[f_2^1/f_1^1], f_1^2(y_2)[f_2^2/f_1^2]) =$$

$$= (\vee E(\text{inj}(x_1^i), f_1^1(y_1), f_1^2(y_2)))[x_2^i, f_2^1, f_2^2/x_1^i, f_1^1, f_1^2] =$$

$$= f_1^i(x_1^i)[x_2^i, f_2^1, f_2^2/x_1^i, f_1^1, f_1^2] = f_1^i(x_1^i)[x_2^i, f_2^i/x_1^i, f_1^i].$$

where $y_1 : A_1$, $y_2 : A_2$, and both are bound by $\vee E$. This may become a definitory condition for justifications, that is, we only allow for justifications $\phi$ such that, for every replacement $\sigma$ of proof-variables in the argument $x$ of $\phi$,

$$\phi(x^\sigma) = (\phi(x))^\sigma.$$

# References

[1] Francez, N. (2015). *Proof-theoretic semantics*. London, College Publications.

[2] Gentzen, G. (1935). Untersuchungen über das logische Schließen I, II, *Mathematische Zeitschrift*, 39, 176–210, 405–431.

[3] Pezlar, I. (2023). Constructive Validity of a Generalized Kreisel-Putnam Rule. https://arxiv.org/abs/2311.15376

[4] Piccolomini d'Aragona, A. (2022). *Prawitz's Epistemic Grounding. An Investigation into the Power of Deduction*. Cham, Springer.

[5] Piecha, T. (2016). Completeness in Proof-Theoretic Semantics. In T. Piecha, P. Schroeder-Heister (eds), *Advances in Proof-Theoretic Semantics*, Cham, Springer, 231–251.

[6] Piecha , T., Schroeder-Heister, P. (2016). Atomic Systems in Proof-Theoretic Semantics: Two Approaches. In J. Redmond, O. Pombo Martin, A. Nepomuceno Fernández (eds), *Epistemology, Knowledge and the Impact of Interaction*, Cham, Springer, 47–62.

[7] Piecha, T., Schroeder-Heister, P. (2019). Incompleteness of Intuitionistic Propositional Logic with respect to Proof-Theoretic Semantics, *Studia Logica*, 107 (1), 233–246.

[8] Prawitz, D. (1965). *Natural Deduction. A Proof-Theoretical Study*. Stockholm, Almqvist & Wiskell.

[9] Prawitz, D. (1973). Towards a Foundation of a General Proof Theory. In P. Suppes, L. Henkin, A. Joja, Gr. C. Moisil (eds), *Proceedings of the Fourth International Congress for Logic, Methodology and Philosophy of Science, Bucharest, 1971*, Amsterdam, Elsevier, 225–250.

[10] Prawitz, D. (2015). Explaining Deductive Inference. In H. Wansing (ed), *Dag Prawitz on Proofs and Meaning*, Cham, Springer, 65–100.

[11] Sandqvist, T. (2015). Base-Extension Semantics for Intuitionistic Sentential Logic, *Logic Journal of the GPL*, 23 (5), 719–731.

[12] Schroeder-Heister, P. (2006). Validity Concepts in Proof-Theoretic Semantics, *Synthese*, 148, 525–571.

[13] Schroeder-Heister, P. (2018). Proof-theoretic Semantics. In E. N. Zalta (ed), *The Stanford Encyclopedia of Philosophy*.

[14] Schroeder-Heister, P. (2024). Prawitz's Completeness Conjecture: a Reassessment, *Theoria*, forthcoming.

[15] Stafford, W. (2021). Proof-Theoretic Semantics and Inquisitive Logic, *Journal of Philosophical Logic*, 50, 1199–1229.

[16] Stafford, W., Nascimento, V. (2023). Following all the Rules: Intuitionistic Completeness for Generalised Proof-Theoretic Validity, *Analysis*.

# THE USE OF EXPERTS IN PROBABILISTIC SEISMIC HAZARD ANALYSIS: TOWARDS A CONFIDENCE APPROACH

LUCA ZANETTI
*Scuola Universitaria Superiore IUSS Pavia*
luca.zanetti@iusspavia.it

DANIELE CHIFFI
*Politecnico di Milano*
daniele.chiffi@polimi.it

LORENZA PETRINI
*Politecnico di Milano*
lorenza.petrini@polimi.it

**Abstract.** Epistemic uncertainties about future earthquakes are included in Probabilistic Seismic Hazard Analysis (PSHA) using logic tree. Each model in the logic tree is assigned a weight based on the judgements provided by a panel of expert. In this paper we raise some methodological challenges to the use of experts in PSHA. We then present an alternative approach to include the scientist's confidence in an estimate for a quantity of interest. The proposed methodology extends Brian Hill's and Richard Bradley's "confidence approach". We finally compare our approach with the standard (Bayesian) approach.

**Keywords:** Probabilistic Seismic Hazard Analysis (PSHA), Logic Tree, Epistemic Uncertainty, Expert Judgement, Confidence Approach.

# 1 Introduction

Probabilistic seismic hazard analysis (PSHA) is the standard methodology for estimating uncertainties due to future earthquakes ([2]). However, historical catalogues of past seismic events, which seldom comprise more than a few centuries, are often insufficient for fully validating probabilistic models, which would require thousands of years of data ([15]). For this reason, estimates of seismic hazard are mostly based on *ensembles* of models, and a key step in this procedure is to form a panel of *experts* that 'weigh' the models. This paper has two goals: (1) to critically discuss some methodological aspects of the use of experts in PSHA, and (2) to present an alternative approach to include the scientist's *confidence* in an estimate for a quantity of interest. The paper proceeds as follows. In Section 2 we introduce the PSHA procedure. In Section 3 we discuss the use of model ensembles to estimate seismic hazards. In Section 4 we apply Brian Hill's and Richard Bradley's "confidence approach" to PSHA. Section 5 concludes the paper.

# 2 Types of Uncertainty in Probabilistic Seismic Hazard Analysis (PSHA)

In PSHA, the seismic hazard of a site is expressed as the probability of exceedance[1] of a specified ground-motion intensity (usually the peak ground acceleration, PGA) at a given site during a specified time interval (for example, a frequency of 10 % in 50 years, corresponding to a return period[2] of 475 years).

The quantification of seismic hazards is subject to two types of uncertainty (see [30] for discussion).

(i) On the one hand, there are *aleatoric* uncertainties, that are due to the essential randomness of seismic phenomena. For example, it is uncertain where future earthquakes will occur (*spatial uncertainty*), when these earthquakes will occur (*temporal uncertainty*), and which level of ground motion they will produce (*ground-motion uncertainty*); see e.g. [4].

Given that aleatoric uncertainty is due to the stochastic nature of seismic processes, it is assumed that this uncertainty does not decrease over time.

The aleatoric variability at a site is represented by the hazard curve. A hazard curve usually plots PGAs (on the $x$-axis) and their frequencies of

---

[1]That is, the probability that each ground-motion intensity level is exceeded within a specified time frame.

[2]The return period is the converse of the frequency.

exceedance (on the $y$-axis). The characteristic shape of the hazard curve displays the fact that earthquakes that produce strong PGAs have lower frequencies (bottom right of the curve) and earthquakes that produce small PGAs have greater frequencies (top left of the curve). Fig. 1 depicts a simplified hazard curve.

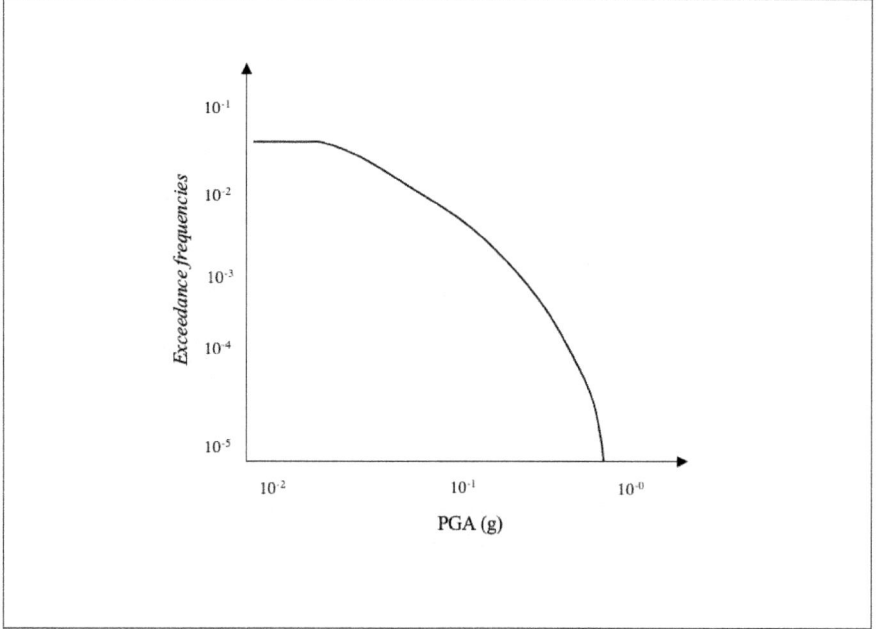

Figure 1: A simplified hazard curve; PGA = peak ground acceleration.

(ii) On the other hand, there are *epistemic* uncertainties, that are due to an insufficient understanding of seismic phenomena or to shortage of historical data ([29]). These uncertainties concern both the general form of the equations that are used to model these phenomena (*model uncertainty*) and the value of the parameters in these equations (*parametric uncertainty*).

Epistemic uncertainties can be represented by a family or *bundle* of hazard curves. Each curve in the bundle corresponds to a model. The bundle itself conveys two pieces of information. Horizontally, the values on the $y$-axis corresponding to the same point in the $x$-axis represent the variance in the estimation of the frequency of an event that produces a given ground-motion intensity. Vertically, the values on the $x$-axis that correspond to the same point in the $y$-axis represent the variance in the estimation of the most intense event with a given frequency.

In principle, epistemic uncertainty is reduced as more data are collected. However, since historical catalogues of strong earthquakes are typically limited, epistemic uncertainties cannot be properly reduced and must therefore be included in the hazard analysis.

## 3 The Bayesian Approach

Epistemic uncertainties are included in PHSA by constructing a *logic tree* ([18]). To set up the logic tree, (1) all the published models that can describe the site (for example, fault-based models and purely statistical models) are included together with the plausible values for their parameters. (2) Each one of these chosen models is then assigned a *weight* expressed as a probability. Fig. 2 depicts a simplified logic tree; the 'final' weights on the rights are multiplied by multiplying the weights of the segments of each brand (e.g., 0.6 and 0.5).

The approach that employs logic trees is Bayesian in two senses[3].

(i) The weights represent the "subjective estimates for the degree-of-certainty or degree-of-belief – expressed within the framework of probability theory – that the corresponding model is the one that should be used" ([25, 1238]). The influential report of the US Senior Seismic Hazard Committee (SSHAC) describes two procedures to formulate the weights of the logic tree ([28, 23]). In the first one, a "technical integrator" (TI) performs a review of the literature, interacts with the proponents of the models, collects the individual judgements of the experts, and finally proposes a probability distribution that represents the scientific community as a whole. The TI has scientific responsibility both for the inputs and the outputs of the PSHA ([28, 31]). In the second procedure, a "technical facilitator/integrator" (TFI) organises a panel of experts, encourages interaction within the panel, and assembles the logic tree with the weights provided directly by the experts. The TFI shares "ownership" of the final result with the other members of the panel.

The report stresses that the goal of this procedure is to reach a *consensus* among the experts ([28, 23, 25-6]). It distinguishes between four types of consensus (from Type 1 to Type 4). What is expected is that all experts should agree that a particular composite probability distribution represents

---

[3]Notice that the interpretation is the same both for the weights that are assigned to models and for the weights that are assigned to the parameters of those models; we are grateful to an anonymous reviewer for stressing this point.

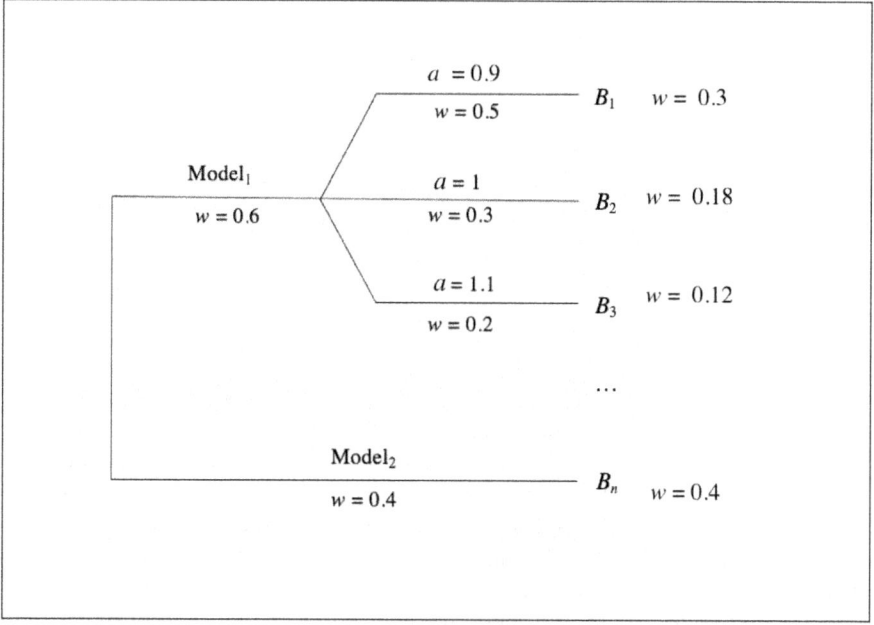

Figure 2: A simplified logic tree; $B_i$ = the $i$-th branch of the logic tree; $w$= weight; $a$= model parameter. The parameters of Model$_2$ are omitted for simplicity.

them as a group (Type 3), or, more weakly, that all experts agree that the distribution represents the overall scientific community (Type 4). It is not required, however, that each expert believes in the same probability distribution for a random variable or model parameter (Type 2) or that each expert agrees with all other experts that a given model is correct or that a given value for a parameter is correct (Type 1).

(ii) Moreover, the weights can be *updated* as new data are collected (this approach was originally proposed by [11], and more recently, by [3] and [26]; see also [1]). Given an observation $A$ consisting of recordings of the seismic activity at the site in a specific interval of time, the probability that the model that corresponds to the $i$-th branch of the logic tree is correct is expressed by Bayes' Law (with $P(A) \neq 0$):

$$P(B_i|A) = \frac{P(B_i)\,P(A|B_i)}{P(A)}, \qquad (1)$$

If the tree has $n$ branches, the prior probability $P(B_i)$ can be calculated using the principle of indifference ([26, 2515]):

$$P(B_i) = \frac{1}{n}, \qquad (2)$$

More often, $P(B_i)$ *is set to the weights assigned to the model $B_i$ by the experts.* Finally, the probability of the observation $A$ is expressed by the total probability theorem,

$$P(A) = \sum_{j=i}^{n} P(B_j)\, P(A|B_j), \qquad (3)$$

as the sum of the probability of observing $A$ if the model $B_j$ is correct weighted by the the probability that $B_j$ is correct.

The posterior probability that a model is correct given some set of observations can be used to score the *predictive power* with respect to an independent set of data conditional on the weights assigned by the experts ([27]). In the literature, this has been seen as a "scientific experiment in the traditional sense", with the hypothesis corresponding to the forecasts of future seismic events and the data consisting of yearly recordings ([20, 444]). More generally, the posterior probability corresponds to the degree of confidence that the experts should have that the model is correct in light of the new data[4].

## 3.1 The precision problem

According to the Bayesian approach, prior probabilities, which correspond in PSHA to the initial weights of the logic trees, have to be *precise*. Therefore, experts must express their epistemic uncertainty using a single probability. In this Section, we point out that this assumption can unrealistic.

First of all, the experts may not have enough *evidence* to form precise judgments. The weights in the logic tree must be grounded in the available empirical evidence: as Küglel emphasises, "subjective probabilities cannot be construed as a license for the expert to say anything whatsoever. Empirical control is a cornerstone of any scientific methodology" ([16, 40]). Lack of evidence is typically mentioned among the reasons to prefer imprecise

---

[4]Notice that this is still rarely done in practice. By contrast, some recent PSHA assigns to models a weight $W = W_1 \cdot W_2$, where $W_1$ is a weight based on the testing phase and $W_2$ is a weight based on experts' judgments [22, 14].

versions of Bayesianism, in which credences are represented by probability intervals rather than by precise probabilities ([8]).

A response that one may give to this problem is that the weights are precise because they correspond to the scientist's estimates of their uncertainty. However, we do not find this answer convincing. As remarked for example by Bradley, Helgeson and Hill, "if the [scientist] has trouble forming precise first-order probabilities, why would he have any less trouble forming precise second-order confidence weights?" ([7, 509]).

Another response is that (precise) probabilities are only a mathematical representation of the epistemic uncertainty of the scientific communities. Experts need not provide a precise representation of their epistemic uncertainty, but it is sufficient that each member of the panel formulates a qualitative assessment of the likelihood of the models. The weights are indeed inferred from these judgments, and the aggregators typically have some choice on how to represent these judgments in the form of a probability distribution ([9]). However, if the experts do not express a probabilistic judgment, the weights assigned to the model are, to some extent, arbitrary.

The problem can be seen, in particular, if we consider the *mean hazard value* is often taken as the reference value for construction. This mean hazard is given by the sum of all the estimates $v_i(a)$ weighted by the weight $w_i$ assigned to the branch $B_i$ of the logic tree. However, if it is arbitrary whether a high level of confidence of the experts corresponds, say, to a weight of .69 or .71, then the method of logic trees can lead either to an underestimation or to an overestimation of the hazard (cf. [19]).

In the next Section, we will present an alternative approach that does *not* require that the individual judgments of the experts are expressed as precise probabilities.

# 4  Going Imprecise: Confidence in PSHA

We will now present an alternative methodology to include the confidence of the scientist in their estimate of the value of a quantity of interest (e.g., the maximum earthquake with a return period of 475 years). Unlike the standard approach in PSHA, the experts are asked to express a *qualitative* level of confidence rather than a precise probability, for ex. "low", "medium" or "high". This approach is similar to the one currently used in climate change modelling, especially in the reports of the Intergovernmental Panel on Climate Change (IPCC). However, it has not been applied in earthquake engineering yet.

Our proposal builds on the "confidence approach" formulated by Brian Hill ([12], [13]) and developed by Richard Bradley ([6]) and by Roussos, Bradley and Frigg ([24]).

Hill proposes that the expert's confidence can be represented as a nested family of sets of probability measures. A probability measure assigns a probability to each member of a set of events (for example, earthquakes that produce a ground-motion intensity that exceeds a specified threshold). Sets of probability measures are often used to represent *imprecise credences*, that is, cases where the expert's degree of belief that $A$ is better represented by a probability interval (intuitively, the interval between the lower probability assigned to $A$ and the highest probability assigned to $A$ in the set) rather than by a precise probability. A nested family of sets is a set that contains a chain of subsets. The nested family of sets represents the expert's levels of confidence and is called a *confidence ranking*. An example of Hill's confidence ranking is represented in Fig. 2.

There are three features of Hill's proposal that are worth mentioning.

(i) Larger sets in the ranking (i.e., outer sets) correspond to higher confidence levels, whereas smaller (i.e., inner) sets correspond to lower levels. Each set in the ranking represents all the values of $Pr(A)$ that the expert holds with the same level of confidence. The expert can have high confidence that the "true" value lies within a given set $\mathcal{A}$, but only medium confidence that it lies in a smaller set $\mathcal{B} \subset \mathcal{A}$, and even less confident that $Pr(A)$ lies in an even smaller set $\mathcal{C} \subset \mathcal{B}$. For example, in Fig. 2 the expert has high confidence that $Pr(A)$ is in the interval between 0.05 and 0.01 since all the estimates in the family of sets are within this interval, but only medium confidence that she can be more precise and say that the probability is in the interval between 0.02 and 0.04.

(ii) Each set in the confidence ranking corresponds to some probability judgements. The set of probability judgements that corresponds to a given set in the confidence ranking is the set of all the judgements that hold for all probability measures in that set. The probability judgements corresponding to a given rank are the judgements the expert holds with the level of confidence represented by that rank. For example, the judgement that $Pr(A) = 0.01$ is included only in the smaller one of the sets represented in Fig. 3. Therefore, the expert believes with low confidence that $Pr(A)$ can be estimated precisely to 0.01.

(ii) Finally, the higher the confidence level we consider, the fewer the probability judgements that the expert holds with that level of confidence since fewer judgements are true for all probability measures in larger sets. The expert holds a given probability judgement with the highest confidence allowed by its rank, namely, with the level corresponding to the larger sets on which that judgement holds. For example, the judgement that $Pr(A) \leq 0.05$ holds in all the sets in the ranking, while $Pr(A) = 0.1$ is outside the confidence rank of the expert. Therefore, the expert is highly confident that $Pr(A)$ is 0.05 or lower.

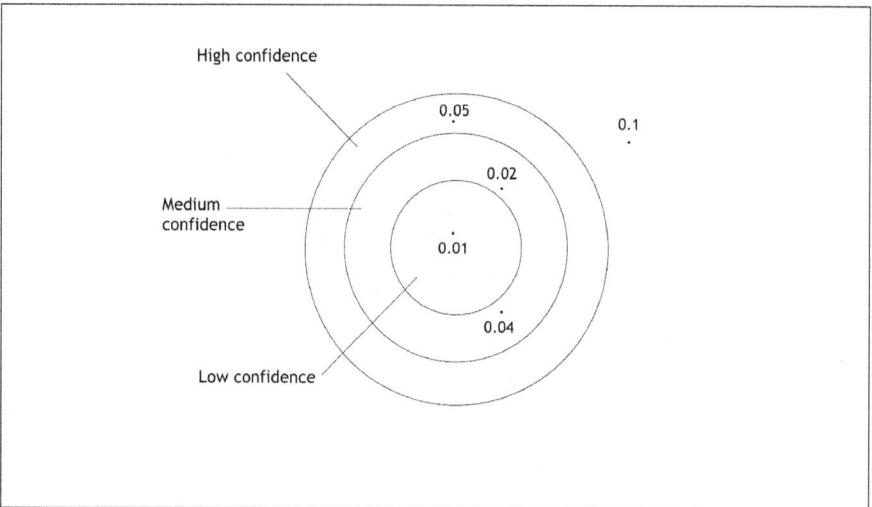

Figure 3: Hill's representation of confidence rankings; the probabilities correspond to different estimates of $Pr(A)$. Each point in the space is a probability assignment.

We will now show how the confidence approach can be used to represent epistemic uncertainty in PSHA. This will require two steps: (1) the definition of a *probability set* containing the estimated frequencies of the same event, and (2) the definition of a *confidence ranking* over this set.

Suppose that we need to estimate the frequency of an event with PGA $> a$, and we want this estimate to include the epistemic uncertainty. We can proceed as follows.

First, the probability set can be defined using the available models. The key idea is that the ensemble of models can be represented as the interval between the lower and the highest value estimated by the logic tree for the same quantity (for example, the frequency of exceedance of a specific PGA). Specifically, every branch of the logic tree can be represented as an *hazard curve* that plots PGAs (on the $x$-axis) and their frequencies of exceedance (on the $y$-axis). The model ensemble can be represented as a *bundle* of *family* of hazard curves. The values on the $y$-axis corresponding to the same point in the $x$-axis (corresponding to a specific PGA) represent the variance in the estimation of the frequency of an event that produces a given ground-motion intensity. Finally, the probability set for PGA $> a$ determined by all the models included in the ensemble can be expressed as the interval $[v(a)_{min}, v(a)_{Max}]$ between the frequency of PGA $> a$ estimated by the 'lowest' curve of the bundle and the frequency estimated by the 'highest' curve[5]

The experts can then provide a confidence ranking over this probability set. This rank should reflect the "weight of evidence" available to them based on the available data and forecastings that are relevant to their judgments ([6, 290]; [7, 506-7]). Importantly, confidence levels are distinct from degrees of credence. Confidence is naturally thought of as an evaluation of the degree of a belief, rather than the degree of belief itself, which is captured by standard probabilities ([13, 227]). For example, new data about frequent but low-intensity seismic events can make the scientist more confident that a particular model is correct even though her estimates remain the same. As summarized by Roussos, Bradley and Frigg, "confidence [reflects] an evaluation of the state of knowledge underpinning [a probabilistic judgment]" ([24, 4]).

Intuitively, the confidence ranking should be centred around the estimate of hazard calculated by the model that receives the highest score. For example, the scientist may have high confidence that "true" frequency lies between the $16^{th}$ and the $84^{th}$ percentiles estimated by the ensemble, but she is less confident in any more precise interval. In particular, the scientist has low confidence that the correct estimate lies precisely around the mean (this is in contrast with the standard procedure in PSHA of taking mean hazards as the "true" hazard values).

However, a particular case can arise when the experts have almost equal confidence in two different intervals of frequencies. For example, in the New Italian Seismic Hazard Maps (MPS19) the two models that received the highest scores from the experts are a seismotectonic model, MA4 ($w = 0.2240$), and a smoothed seismicity model, MS1 ($w = 0.2193$). Even though in the confidence approach we do *not* assign weights to the models, we assume that the experts in the panel of MPS19 would place equal confidence in the estimates of MA4 and MS1 for the same quantity.

---

[5]Alternatively, this set may contain only the specific estimates calculated by the models rather than the interval between the lowest and the highest estimates of the ensemble.

In particular, the mean annual frequency of an earthquake with magnitude $M \geq 4.5$ in the Italian region, based on all the eleven models that were included in MPS19, is between $\sim 4/\text{yr}$ and $\sim 8/\text{yr}$. However, the average frequency according to MS1 is 4.75 and the average frequency according to MA4 is 6.1/year ([22]). Therefore, a scientist who is confident that either MA4 or MS1 is correct should hold with high confidence that the frequency of $M \geq 4.5$ is either $\sim 4.75/\text{yr}$ or $\sim 6.1/\text{yr}$, with low confidence in values far off from both those estimates.

These confidence judgements can be represented by modifying Hill's original approach by allowing for nested families of sets with multiple *foci*. The fact that the confidence ranking has multiple *foci* does not mean that the expert thinks that there are multiple "true" frequencies of the same event, but only that the evidence available to the expert is such that it permits to hold two disjoint probability intervals with equal confidence. For example, Fig. 4 represents the confidence of the experts in the panel of MPS19 for the frequency of an earthquake with $M \geq 4.5$ in the Italian region.

It is important to notice that the fact that the experts gave high weights to the two models does not guarantee that these experts would automatically have high confidence in the estimates of frequency calculated by these models. The goal of using this alternative procedure is indeed to elicit how much the experts *trust the final estimate* rather than to quantify the degree of confidence of these experts that the model is correct. Moreover, Fig. 4 represents the confidence of the entire panel of experts; a single expert may believe that only one of the two models is correct, and, therefore, have high confidence in a single interval.

The confidence approach differs from the Bayesian approach that is currently prevailing in earthquake engineering (cf. Sect. 3) in the following ways:

(a) PSHA focuses on a single estimate of seismic hazard (either the mean value or a percentile), whereas the approach presented in this paper yields probability intervals (that are held with a specified level of confidence).

(b) This approach treats the confidence of the experts in a specif probability judgment as an *ordinal* notion, whereas standard PSHA treats it as a cardinal notion (as probability weights in the logic tree).

(c) The standard methodology is 'bottom up': the inputs of the PSHA consist of some sets of historical data from the area in which the models will be applied. The output consists of the final estimate of

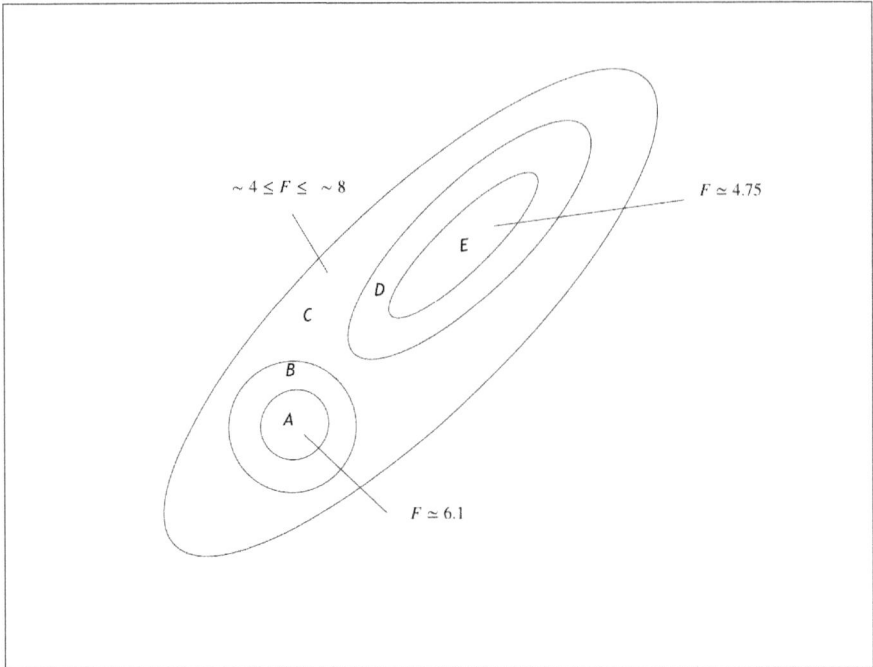

Figure 4: Confidence ranking with two *foci* for the frequency $F$ of an earthquake with $M \geq 4.5$ based on MPS19; $A$ and $E$ = low confidence, $B$ and $D$ = medium confidence, $C$ = high confidence.

seismic hazard, given in the form of a single probability (e.g., the mean value of the logic tree or the $85^{th}$ or $90^{th}$ percentile); by contrast, the confidence approach is 'top down': experts assess directly their confidence about the estimate of a single quantity (for example, the frequency of an event with a ground-motion intensity that exceeds a specified threshold) based on the available historical data and the models that have been proposed in the literature.

(d) Finally, in the credibility approach there is a clear distinction between the estimates of seismic hazard (produced by the models) and the confidence ranking given by the experts. The second approach is therefore clearer and more usable than the standard approach by the decision-makers, who receive different confidence levels from the experts rather than a single value (mean hazard or percentiles).

## 5 Conclusions: Towards Decision-Making

In this paper, we critically discussed a methodological issue with the use of logic trees in probabilistic seismic hazard analysis, namely the 'precision' challenge. This challenge is due to the fact that the weights in the logic tree must be precise, but historical data are not sufficient to fully validate seismic hazard models. We also presented an alternative approach to include the scientist's confidence in an estimate for a quantity of interest.

This approach is promising for three reasons. First, it distinguishes between the weight of evidence in favour of a probability judgement, encoded in the confidence level, and the probability judgement itself. For example, new data about frequent but low-intensity seismic events can make the scientist more confident that a particular model is correct even though the scientist's best estimate of the frequency of those events remains the same. Second, confidence levels are distinct from degrees of credence. This reflects the current practice of eliciting experts' individual judgements as qualitative assessments rather than directly as probabilities. Third and finally, the confidence in a probability judgement required for that judgement to play a role in a decision depends on what is at stake in that decision. This corresponds to the way in which PSHA is currently practised, since the assessment of seismic hazard models may be influenced by practical goals and aims such as safety ([10]), and do not depend merely on the scientific accuracy of the models themselves ([30, 31]).

The confidence approach is not likely to improve the predictions of seismic hazard. However, it does provide a better representation of the scientific uncertainty surrounding this predictions that the standard (Bayesian) approach. Finally, the proponents of the confidence approach stress, for example, that "the larger the stakes involved in a decision, the more confidence is required in a probability judgement for it to play a role in the decision". High-stakes decisions (for example, building a power plant) may call for high confidence, whereas medium or low confidence may suffice in low-stakes situations (for example, building a deposit in an isolated area). Once that the appropriate level of confidence is selected, however, the approach is compatible with multiple decision rules with imprecise probabilities. This flexibility is typically not available in the standard approach, which mixes together the estimates of the models and the judgements of the experts into the definition of the mean hazard. By contrast, the confidence approach distinguishes between these two dimensions of uncertainty (aleatory and epistemic) and makes both of them available to decision-makers.

## Acknowledgments

An early version of this paper received the Best Presentation Award of the Italian Society for Logic and the Philosophy of Science (SILFS) during the 5h SILFS Postgraduate Conference (Milan-Bicocca, 13-17 June 2022). This paper was presented during the CS3 Seminar *Environmental, Natural and Antrophic Risks: From Modelling to Ethical Decisions* at the Dept. of Civil and Environmental Engineering (DICA) of Politecnico di Milano (Milan, 3 November 2022); we wish to thank Richard Bradley and Roberto Paolucci for their helpful comments. We are finally grateful to two anonymous reviewers for their comments, which contributed to improving this paper. This research was partially funded by Next Generation EU, Piano Nazionale di Ripresa e Resilienza (PNRR), Ministry of University and Research: "RETURN. Multi-Risk Science for Resilient Communities Under a Changing Climate".

## References

[1] Alamilla, J. L.; Rodriguez, J. A., and Vai, R. (2020). Unification of Different Approaches to Probabilistic Seismic Hazard Analysis. *Bulletin of the Seismological Society of America* 110(6): 2816–2827.

[2] Baker, J., Bradley, B., Stafford, P. (2021). *Seismic Hazard and Risk Analysis*. Cambridge: Cambridge University Press.

[3] Baker, J., Gupta, A. (2016). Bayesian Treatment of Induced Seismicity in Probabilistic Seismic-Hazard Analysis. *Bulletin of the Seismological Society of America* 106: 1-11.

[4] Bazzurro, P., Cornell, C.. (1999). Disaggregation of Seismic Hazard. *Bulletin of the Seismological Society of America* 89(10): 501–520.

[5] Bommer, J. J. and Scherbaumb, F. (2008). The Use and Misuse of Logic Trees in Probabilistic Seismic Hazard Analysis. *Earthquake Spectra* 4(24): 997–1009.

[6] Bradley, R. (2017). *Decision Theory With a Human Face*. Cambridge University Press.

[7] Bradley, R., Helgeson, C., Hill, B. (2017). Climate Change Assessments: Confidence, Probability and Decision. *Philosophy of Science*, 84: 500-22.

[8] Bradley, S. (2019). Imprecise Probabilities, in Zalta, E. (Ed.), *The Stanford Encyclopedia of Philosophy*, Spring 2019 Edition.

[9] Budnitz, R. J., Apostolakis, G., Boore, D.M., Cluff, L.S., Coppersmith, K.J., Cornell, C.A., Morris, P.A. (1998). Use of Technical Expert Panels: Applications to Probabilistic Seismic Hazard Analysis. *Risk Analysis* 18(4): 463-9.

[10] Diekmann, S., Peterson, M. (2013). The Role of Non-Epistemic Values in Engineering Models. *Science and Engineering Ethics*, 19(1): 207–218.

[11] Esteva, L. (1969). Seismicity prediction: A Bayesian Approach, *Proceedings of the Fourth World Conference on Earthquake Engineering*, Santiago, Chile.

[12] Hill, B. (2013). Confidence and Decision. *Games and Economic Behavior* 82: 675-92.

[13] Hill, B. (2019). Confidence in Beliefs and Rational Decision Making. *Economics and Philosophy* 35(2): 223-58.

[14] IPCC – Intergovernmental Panel on Climate Change (2013). *Climate Change 2013: The Physical Science Basis. Contribution of Working Group I to the Fifth Assessment Report of the Intergovernmental Panel on Climate Change.* Cambridge: Cambridge University Press.

[15] Kijko, A. (2011). Seismic Hazard. In Gupta, H. (Ed.). *Encyclopedia of Solid Earth Geophysics*: 1107-20. Dordrecht: Springer.

[16] Klügel, J. (2011). Uncertainty Analysis and Expert Judgment in Seismic Hazard Analysis. *Pure and Applied Geophysics* 168: 27-53.

[17] Krinitzsky, E. L. (2002). Epistematic and Aleatory Uncertainty: A New Shtick for Probabilistic Seismic Hazard Analysis. *Environmental & Engineering Geoscience* IV(4): 425-43.

[18] Kulkarni, R. B.; Youngs, R. R. and Coppersmith, K. J. (1984). Assessment of Confidence Intervals for Results of Seismic Hazard Analysis, *Proceedings, Eighth World Conference on Earthquake Engineering* 1: 263–270.

[19] Marulanda, M. C., de la Llera, J. C., Bernal, G. A., Cardona, O. D. (2021). Epistemic Uncertainty in Probabilistic Estimates of Seismic Risk Resulting from Multiple Hazard Models. *Natural Hazards* 108: 3203–3227.

[20] Marzocchi, W., Zechar, J. D. (2011). Earthquake Forecasting and Earthquake Prediction: Different Approaches for Obtaining the Best Model. *Seismological Research Letter* 82(3): 442-8.

[21] McGuire, R., Cornell, C. A., Toro, G. (2005). The Case for Using Mean Seismic Hazard. *Earthquake Spectra* 21(3): 879–886.

[22] Meletti, C., Marzocchi, W., D'Amico, V., Lanzano, G., Luzi, L., Martinelli, F., Pace, B., Rovida, A., Taroni, M., Visini, F. (2021). "The new Italian Seismic Hazard Model (MPS19)". *Annals of Geophysics*, 64 (1), DOI: 10.4401/ag-8579.

[23] Musson, R. (2012). On the Nature of Logic Trees in Probabilistic Seismic Hazard Assessment. *Earthquake Spectra* 28: 1291–1296.

[24] Roussos, J., Bradley, R., Frigg, R. (2021). Making Confident Decisions with Model Ensembles. *Philosophy of Science* 88 (3): 439-460.

[25] Scherbaum, F., Kuehn, N. M. (2011). Logic Tree Branch Weights and Probabilities: Summing Up to One is not Enough. *Earthquake Spectra* 27: 1237–1251.

[26] Secanell, R.; Martin, C.; Viallet, E. and Senfaute, G. (2018). "A Bayesian Methodology to Update the Probabilistic Seismic Hazard Assessment". *Bulletin Earthquake Engineering* 16: 2513–2527.

[27] Selva, J., Sandri, L. (2013). Probabilistic Seismic Hazard Assessment: Combining Cornell-Like Approaches and Data at Sites through Bayesian Inference. *The Bulletin of the Seismological Society of America* 103: 1709-1722.

[28] SSHAC – Senior Seismic Hazard Analysis Committee (1997). *Recommendations for Probabilistic Seismic Hazard Analysis: Guidance on Uncertainty and Use of Experts.* Report NUREG-CR-6372, U.S. Nuclear Regulatory Commission, Washington D.C.

[29] Wang, Z., Woolery, E. W., Shi, B., Kiefer, J. D. (2003). Communicating with Uncertainty: A Critical Issue with Probabilistic Seismic Hazard Analysis. *Eos* 84(46): 501, 506-508.

[30] Zanetti, L., Chiffi, D., Petrini, L. (2023a). Philosophical Aspects of Probabilistic Seismic Hazard Analysis. *Natural Hazards* 117, 1193–1212.

[31] Zanetti, L., Chiffi, D., Petrini, L. (2023b). Epistemic and Non-Epistemic Values in Earthquake Engineering. *Science and Engineering Ethics* 29(3), 1-16.

www.ingramcontent.com/pod-product-compliance
Lightning Source LLC
Chambersburg PA
CBHW071426160426
43195CB00013B/1827